Lecture Notes in Computer Science 4240

Commenced Publication in 1973
Founding and Former Series Editors:
Gerhard Goos, Juris Hartmanis, and Jan van Leeuwen

T0223354

Sotiris E. Nikoletseas   José D.P. Rolim (Eds.)

Algorithmic Aspects of Wireless Sensor Networks

Second International Workshop, ALGOSENSORS 2006
Venice, Italy, July 15, 2006
Revised Selected Papers

 Springer

Volume Editors

Sotiris E. Nikoletseas
University of Patras and Computer Technology Institute (CTI)
N. Kazantzaki Street, 26500, Rio, Patras, Greece
E-mail: nikole@cti.gr

José D.P. Rolim
Centre Universitaire d'Informatique
Battelle bâtiment A, route de Drize 7, 1227 Carouge, Geneva, Switzerland
E-mail: rolim@cui.unige.ch

Library of Congress Control Number: 2006938988

CR Subject Classification (1998): F.2, C.2, E.1, G.2

LNCS Sublibrary: SL 5 – Computer Communication Networks and
Telecommunications

ISSN   0302-9743
ISBN-10   3-540-69085-9 Springer Berlin Heidelberg New York
ISBN-13   978-3-540-69085-6 Springer Berlin Heidelberg New York

Springer is a part of Springer Science+Business Media

springer.com

© Springer-Verlag Berlin Heidelberg 2006
Printed in Germany

Typesetting: Camera-ready by author, data conversion by Scientific Publishing Services, Chennai, India
Printed on acid-free paper   SPIN: 11963271   06/3142   5 4 3 2 1 0

Preface

This volume contains the papers presented at the 2nd International Workshop on Algorithmic Aspects of Wireless Sensor Networks (ALGOSENSORS 2006), held July 15, 2006, in Venice, Italy, co-located with the 33rd International Colloquium on Automata, Languages, and Programming (ICALP 2006).

The ALGOSENSORS event series aims at reinforcing the foundational and algorithmic aspects of wireless sensor networks research. In particular, ALGO-SENSORS focuses on abstract models, complexity theoretic results and lower bounds, as well as the design and analysis of algorithms for wireless sensor networks.

This volume contains the 20 papers (15 regular and 5 short) that were selected after a rigorous review process by the Program Committee (PC) from 68 submitted papers. Each paper was reviewed by at least 2 PC members, while many papers were reviewed by 3 PC members. The broad PC was composed of 29 leading researchers worldwide, covering several aspects of this research area.

Comments by the PC were provided to the authors of all submitted papers. Furthermore, this year the proceedings were prepared after the event. In this way, authors had the opportunity to revise their papers in light of the discussion at the event and then submit the final versions included in this volume.

The contributed papers span several important research issues, including network topology aspects (graph models, connectivity, partitioning), deployment strategies, distributed computing issues (self-stabilization, initialization), localization and tracking problems, interference minimization, security aspects, broadcasting and communication, energy management.

Selected papers of ALGOSENSORS 2006 will be considered for publication in a Special Issue on Algorithmic Aspects of Wireless Sensor Networks of the *Theoretical Computer Science (TCS) Journal*, edited by S. Nikoletseas and J. Rolim.

Although having an international character (as also witnessed by the composition of the PC and the geographic diversity of the submitted and accepted papers), ALGOSENSORS has a strong European perspective; in particular, we greatly appreciate the relationship with the European Association of Theoretical Computer Science (EATCS).

In order to better design and coordinate the future strategy of the ALGO-SENSORS series of events, a Steering Committee was formed, composed of Josep Diaz (T.U. of Catalonia, Spain), Jan van Leeuwen (U. of Utrecht, The Netherlands), Sotiris Nikoletseas (U. of Patras and CTI, Greece, Chair), José Rolim (U. of Geneva, Switzerland) and Paul Spirakis (U. of Patras and CTI, Greece).

First, we like to warmly thank P.R. Kumar (U. of Illinois, Urbana-Champaign, USA) for delivering a very interesting Invited Talk titled "Computation, Timing and Control in Sensor Networks." We would like to thank all the authors who submitted papers to ALGOSENSORS 2006, the members of the PC, as well as the trusted external referees. Also we thank the members of the Organizing Committee. In particular, we thank the Publicity Chair Ioannis Chatzigiannakis for an excellent job and Athanasios Kinalis (M.Sc.) for maintaining the Web page and efficiently integrating this proceedings volume.

We gratefully acknowledge the support from the Research Academic Computer Technology Institute (RACTI, Greece) and the TCSensor Lab of the University of Geneva (Switzerland). We thank the Athens Information Technology (AIT, Greece) Center of Excellence for Research and Graduate Education and Intracom Defense Electronics for their support. We also thank the European Union (EU) IST/FET ("Future and Emerging Technologies") R&D Project of the Global Computing (GC) Proactive Initiative AEOLUS (Integrated Project IST-15964, "Algorithmic Principles for Building Efficient Overlay Computers"). Finally, we wish to thank Springer's *Lecture Notes in Computer Science* (LNCS) team for a very nice and efficient cooperation.

October 2006   Sotiris Nikoletseas and José Rolim
                                                ALGOSENSORS 2006 PC Co-chairs

Organization

Program Committee Co-chairs

Sotiris Nikoletseas	University of Patras and CTI, Greece
José Rolim	University of Geneva, Switzerland

Program Committee

Ian Akyildiz	Georgia Institute of Technology, USA
Azzedine Boukerche	University of Ottawa, Canada
Costas Busch	Rensselaer Polytechnic Institute, USA
Ioannis Chatzigiannakis	University of Patras and CTI, Greece
Andrea Clementi	University of Rome "Tor Vergata," Italy
Josep Diaz	Technical University of Catalonia, Spain
Tassos Dimitriou	Athens Information Technology, Greece
Shlomi Dolev	Ben-Gurion University, Israel
Deborah Estrin	UCLA, USA
Alfredo Ferro	University of Catania, Italy
Stefan Fischer	University of Luebeck, Germany
Pierre Fraigniaud	CNRS, University Paris-Sud, France
Jorge Garcia-Vidal	Technical University of Catalonia, Spain
Chalermek Intanagonwiwat	Chulalongkorn University, Thailand
Christos Kaklamanis	University of Patras and CTI, Greece
Miroslaw Kutylowski	Wroclaw University of Technology, Poland
Jan van Leeuwen	University of Utrecht, The Netherlands
Alberto Marchetti Spaccamela	University of Rome "La Sapienza," Italy
Sotiris Nikoletseas (Co-chair)	University of Patras and CTI, Greece
Stephan Olariu	Old Dominion University, USA
Pekka Orponen	Helsinki University of Technology, Finland
Pino Persiano	University of Salerno, Italy
José Rolim (Co-chair)	University of Geneva, Switzerland
Christian Schindelhauer	University of Freiburg, Germany
Paul Spirakis	University of Patras and CTI, Greece
Philippas Tsigas	Chalmers University, Sweden
Peter Widmayer	ETH Zurich, Switzerland
Jiri Wiedermann	Academy of Sciences, Czech Republic
Manos Varvarigos	University of Patras and CTI, Greece

Organizing Committee

Ioannis Chatzigiannakis	University of Patras and CTI, Greece (Chair)
Athanasios Kinalis	University of Patras and CTI, Greece
Giorgos Mylonas	University of Patras and CTI, Greece

Referees

Beccehtti, Luca
Calamoneri, Tiziana
Caragiannis, Ioannis
Crescenzi, Pilu
Ferro, Alfredo
Giugno, Rosalba
Gkamas, Vassilis
Gomułkiewicz, Marcin
Jarry, Aubin
Kanellopoulos, Panagiotis
Kinalis, Athanasios

Klonowski, Marek
Kokkinos, Panagiotis
Krzywiecki, Łukasz
Lauks, Anna
Leone, Pierre
Marculescu, Andrei
Moraru, Luminita
Moscardelli, Luca
Mylonas, Giorgos
Papageorgiou, Christos
Papaioannou, Evi

Pasquale, Francesco
Powell, Olivier
Pulvirenti, Alfredo
Silvestri, Riccardo
Vitaletti, Andrea
Zagórski, Filip
Zawada, Marcin

Sponsoring Institutions

- Research Academic Computer Technology Institute (R.A.C.T.I.), Greece.
- The TCSensor Lab of the University of Geneva, Switzerland.
- EU-FET R&D Project "Algorithmic Principles for Building Efficient Overlay Computers" (AEOLUS).
- Intracom Defense Electronics, Greece.
- Athens Information Technology (AIT), Greece.

Table of Contents

Regular Papers

Short Papers

Efficient Training of Sensor Networks

A.A. Bertossi[1], S. Olariu[2], and M.C. Pinotti[3]

[1] Department of Computer Science, University of Bologna,
Mura Anteo Zamboni 7, 40127 Bologna, Italy
`bertossi@cs.unibo.it`
[2] Department of Computer Science, Old Dominion University,
Norfolk, VA 23529-0162, USA
`olariu@cs.odu.edu`
[3] Department of Computer Science and Mathematics,
University of Perugia, 06123 Perugia, Italy
`pinotti@unipg.it`

Abstract. Due to their small form factor and modest energy budget, individual sensors are not expected to be GPS-enabled. Moreover, in most
applications, exact geographic location is not necessary, and all that the
individual sensors need is a coarse-grain location awareness. The task of
acquiring such a coarse-grain location awareness is referred to as training.
In this paper, a scalable energy-efficient training protocol is proposed for
massively-deployed sensor networks, where sensors are initially anonymous and unaware of their location. The training protocol is lightweight
and simple to implement; it is based on an intuitive coordinate system
imposed onto the deployment area which partitions the anonymous sensors into clusters where data can be gathered from the environment and
synthesized under local control.

1 Introduction

Recent advances in nano-technology have made it feasible to develop miniatur-
ized low-power devices that integrate sensing, special-purpose computing and
wireless communications capabilities [1,12,19]. These small devices, commonly
called *sensors*, will be mass-produced, making their production cost negligible. A
sensor has a small, non-renewable power supply and, once deployed, must work
unattended. A massive deployment of sensors, perhaps in the order of thousands
or even tens of thousands [17], is expected.

Aggregating sensors into sophisticated computational and communication infrastructures, called *wireless sensor networks*, will have a significant impact on
a wide array of applications ranging from military, to scientific, to industrial, to
health-care, to domestic, establishing ubiquitous wireless sensor networks that
will pervade society redefining the way in which we live and work [12]. The nov-
elty of wireless sensor networks and the tremendous potential for a multitude of
application domains has triggered a lot of activity in both academia and industry
[3,4,5,7,9].

S. Nikoletseas and J.D.P. Rolim (Eds.): ALGOSENSORS 2006, LNCS 4240, pp. 1–12, 2006.

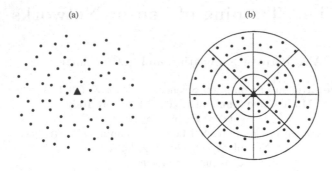

Fig. 1. (a) A sensor network with a central sink node. (b) The trained sensor network.

The peculiar characteristics of sensor networks (a massive deployment of sensors, the anonymity of individual sensors, a limited battery power budget per sensor, and a possibly hostile environment) pose unique challenges to the design of protocols. First of all, the limited power budget requires the design of ultra-lightweight communication protocols. However, how data collected by sensors are queried and accessed and how concurrent sensing can be performed internally are of significance as well. An important guideline in this direction is to perform as much local data processing at the sensor level as possible, avoiding the transmission of raw data through the sensor network. This implies that the sensor network must be multi-hop and only a small number of sensors have the sink as one of their one-hop neighbors. For reasons of scalability, it is assumed that no sensor knows the topology of the network.

Several possible techniques can be used for interfacing sensor networks to the outside world and for harvesting the data they produce. The simplest technique involves using one or several sink nodes, i.e. special long-range radios deployed alongside with the sensors. Each sink has a full range of computational capabilities, can send long-range directional broadcasts to the sensors at distance at most R, can receive messages from nearby sensors, and has a steady power supply. In this scenario, the raw data collected by individual sensors are fused, in stages, and forwarded to the nearest sink that provides the interface to the outside world. Such a scenario for a sensor network with a single central sink is depicted in Figure 1(a).

There are some applications requiring sensory data with exact geographical location, motivating the development of communication protocols that are location aware and perhaps location dependent. In some other applications, however, exact geographic location is not necessary, and all that the individual sensors need is only coarse-grain location awareness [11,16]. Of course, there is a trade-off, because coarse-grain location awareness is lightweight but the resulting accuracy is only a rough approximation of the exact geographic coordinates.

The random deployment results in sensors initially unaware of their location. Further, due to limitations in form factor, cost per unit and energy budget, individual sensors are not expected to be GPS-enabled. Moreover, many probable application environments limit satellite access. Therefore, individual sensors have to determine their exact geographic location, if required by the application, or else a coarse-grain approximation thereof. The former task is referred to as *localization* and has been extensively studied in the literature [8,10]. The latter task, referred to as *training*, has been considered by Olariu et al. [11,16,18]. In particular, they devised two elegant training protocols for sensor networks, which differ on whether or not sensors need some kind of synchronization with the sink. Such two training protocols have different performance, measured in terms of overall time for training, sensor awake time, and number of sensor wake/sleep transitions.

The main contribution of the present paper is to present a new training protocol which outperforms that originally presented in [16], lowering its overall time for training from a linear to a square-root function of the size of the coordinate system used for location awareness. The protocol assumes that the sink and the sensors are somehow synchronized [11,13,14].

The remainder of this paper is organized as follows. Section 2 discusses the sensor model used throughout the work. Section 3 introduces the task of training, that is, endowing individual sensors with coarse-grain location awareness. Training imposes a coordinate system which divides the sensor network area into equiangular wedges and concentric coronas centered at the sink, as first suggested in [16]. Section 4 is the backbone of the entire paper, presenting the theoretical underpinnings of the training protocol. In the corona training protocol, time is ruled into slots and each sensor has to learn a string of bits representing its corona number. The protocol consists of two phases. The first phase is centralized and sink-driven. Its computation can be thought as a visit of a complete binary tree, whose leaves represent coronas, whose vertex preorder numbers are related to the time slots, and whose vertex inorder numbers are related to the transmission range used by the sink. At the end of the first phase, sensors that belong to a group of some consecutive coronas have learned the same most significant bits. The second phase is distributed and, within each group, the sensors that have already known their corona number inform those in the next corona to properly set their remaining bits. Finally, Section 5 offers concluding remarks.

2   The Sensor Model

We assume a sensor to be a device that possesses three basic capabilities: sensory, computation, and wireless communication. The sensory capability is necessary to acquire data from the environment; the computational capability is necessary for aggregating data, processing control information, and managing both sensory and communication activity. Finally, the wireless communication capability is necessary for sending/receiving aggregated data and control information to/from other sensors or the sink.

We assume that individual sensors are tiny, mass-produced devices that operate subject to the following fundamental constraints.

a. Sensors are *anonymous* – they do not have individually unique IDs;
b. Each sensor has a modest non-renewable energy budget;

c. In order to save energy, sensors are in *sleep* mode most of the time, waking up for short intervals;
d. Each sensor has a modest transmission range, perhaps a few meters – this implies that out-bound messages can reach only the sensors in its proximity, typically a small fraction of the sensors deployed;
e. Individual sensors must work *unattended* – once deployed it is either infeasible or impractical to devote attention to individual sensors.

It is worth mentioning that while the energy budget can supply short-term applications, sensors dedicated to work over extended periods of time may need to scavenge energy from the specific environment they are placed into, employing light, temperature, vibration, kinetics, magnetic fields, etc.

3   Training a Sensor Network

In this work we assume a wireless sensor network that consists of a sink and a set of sensors randomly deployed in its broadcast range as illustrated in Figure 1(a). For simplicity, we assume that the sink is centrally placed, although this is not really necessary.

The task of endowing sensors with coarse-grain location awareness, referred to as training, is essential in several applications. One example is *clustering* where the set of sensors deployed in an area is partitioned into clusters [1,2,6,15]. As a result of training, we impose a coordinate system onto the sensor network in such a way that each sensor belongs to exactly one cluster. The coordinate system divides the sensor network area into equiangular wedges. In turn, these wedges are divided into sectors by means of concentric circles or coronas centered at the sink and whose radii are determined to optimize the transmission efficiency of sensors-to-sink transmission. Sensors in a given sector map to a cluster, the mapping between clusters and sectors being one-to-one. In particular, a cluster is the locus of all nodes having the same coordinates in the coordinate systems [11].

Referring to Figure 1(b), the task of training a sensor network involves establishing [16]:

1. *Coronas*: The deployment area is covered by k coronas determined by k concentric radii $0 < r_1 < r_2 < \cdots < r_k = R$ centered at the sink.
2. *Wedges*: The deployment area is ruled into a number of angular wedges, centered at the sink, which are established by directional transmission [11].

As illustrated in Figure 1(b), at the end of the training period each sensor has acquired two coordinates: the identity of the corona in which it lies, as well as the identity of the wedge to which it belongs. Importantly, the locus of all the sensors that have the same coordinates determines a cluster.

4 The Corona Training Protocol

The main goal of this subsection is to present the details of the corona training protocol. The wedge training protocol is similar (in fact, simpler) and will not be discussed.

Let k be an integer known to the sensors and let the k coronas be determined by concentric circles of radii $r_1 < r_2 < \cdots < r_k$ centered at the sink node. For simplicity we shall assume that k is a power of two. Given the transmission range r of each sensor, the radius r_i is assumed to be equal to ir, namely the corona width is $r_{i+1} - r_i = r$.

The idea of the corona training protocol is for each individual sensor to learn the identity of the corona to which it belongs. For this purpose, each individual sensor learns a string of $\log k$ bits from which the corona number can be determined easily. To see how this is done, it is useful to assume time ruled into slots and that the sensors can synchronize to the master clock running at the sink node.

In time slot s_1 all the sensors are awake and the sink transmits with a power level corresponding to $r_{\frac{k}{2}}$. In other words, in the first slot the sensors in the first $\frac{k}{2}$ coronas will receive the message above a certain threshold, while the others will not. Accordingly, the sensors that receive the signal set $b_1 = 0$, the others set $b_1 = 1$.

Consider a k-leaf complete binary tree T, whose leaves are numbered left to right from 1 to k. The edges of T are labeled by 0's and 1's in such a way that

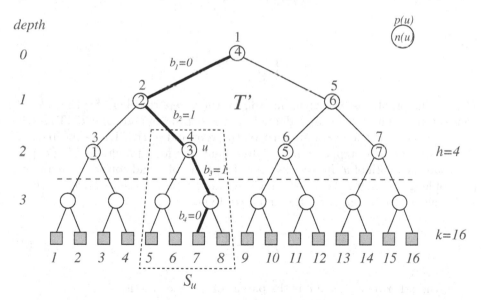

Fig. 2. Illustrating corona training. Labels outside nodes of T' give their preorder numbers, while labels inside give their inorder numbers. Leaves represent coronas, numbered from 1 to k.

an edge leading to a left-subtree is labeled by a 0 and an edge leading to a right subtree is labeled by a 1. Let ℓ, $(1 \leq \ell \leq k)$, be an arbitrary leaf and let $b_1, b_2, \ldots, b_{\log k}$ be the edge labels of the unique path leading from the root to ℓ. It is both well known and easy to prove by a standard inductive argument that

$$\ell = 1 + \sum_{j=1}^{\log k} b_j 2^{\log k - j}. \tag{1}$$

For example, refer to Figure 2, where $k = 16$. Applying Equation (1) to leaf 7, we have: $7 = 1 + 0 * 2^3 + 1 * 2^2 + 1 * 2^1 + 0 * 2^0$.

Let h be an integer known to the sensors which is a power of two such that $1 \leq h \leq k/2$. Consider the subtree T' consisting of the uppermost $2h - 1$ nodes of T. Refer again to Figure 2, where $h = 4$ and T' consists of the uppermost 7 nodes. Let u be an arbitrary node in T', other than the root, and let $b_1, b_2, \ldots, b_i$ be the edge labels on the unique path from the root to u, where i is the depth of u in T' and $1 \leq i \leq \log h$. Note that the root of T' is at depth $i = 0$, and it is characterized by an empty sequence of edge labels.

We take note of the following technical result.

Lemma 1. *Let u be an arbitrary node of depth i in T'. Then, the preorder number $p(u)$ of u is given by*

$$p(u) = 1 + \sum_{j=1}^{i} c_j$$

where

$$c_j = \begin{cases} 1 & \text{if } b_j = 0 \\ \frac{h}{2^{j-1}} & \text{if } b_j = 1. \end{cases}$$

Proof. The proof was first given in [16], and it is reported here for the sake of completeness. The proof is by induction on the depth i of node u in T'. To settle the basis, note that for $i = 0$, u must be the root and $p(u) = 1$, as expected.

For the inductive step, assume the statement true for all nodes in T' of depth less than the depth of u. Indeed, let v be the parent of u and consider the unique path of length i joining the root to u. Clearly, nodes u and v share $b_1, b_2, \ldots, b_{i-1}$ and, thus, $c_1, c_2, \ldots, c_{i-1}$. By the inductive hypothesis,

$$p(v) = 1 + \sum_{j=1}^{i-1} c_j. \tag{2}$$

On the other hand, since v is the parent of u, we can write

$$p(u) = p(v) + \begin{cases} 1 & \text{if } u \text{ is the left child of } v \\ \frac{h}{2^{i-1}} & \text{otherwise} \end{cases} \tag{3}$$

Notice that if u is the left child of v we have $b_i = 0$ and $c_i = 1$; otherwise $b_i = 1$ and $c_i = \frac{h}{2^{i-1}}$. This observation, along with (2) and (3) combined, allows us to write

$$p(u) = 1 + \sum_{j=1}^{i-1} c_j + c_i = 1 + \sum_{j=1}^{i} c_j$$

completing the proof of the lemma.

As an example, consider node u in Figure 2. Applying Lemma 1, one gets $p(u) = 1 + 1 + \frac{4}{2^1} = 4$.

Lemma 2. *Let u be an arbitrary node of depth i in T'. Then, the inorder number $n(u)$ of u is given by*

$$n(u) = h + \sum_{j=1}^{i} d_j$$

where

$$d_j = \begin{cases} -\frac{h}{2^j} & \text{if } b_j = 0 \\ +\frac{h}{2^j} & \text{if } b_j = 1 \end{cases}$$

Proof. The proof is by induction on the depth i of node u in T'. To settle the basis, note that for $i = 0$, u must be the root and $n(u) = h$, as expected.

For the inductive step, assume the statement true for all nodes in T' of depth less than the depth of u. Indeed, let v be the parent of u and consider the unique path of length i joining the root to u. Clearly, nodes u and v share $b_1, b_2, \ldots, b_{i-1}$ and, thus, $d_1, d_2, \ldots, d_{i-1}$. By the inductive hypothesis,

$$n(v) = h + \sum_{j=1}^{i-1} d_j. \tag{4}$$

On the other hand, since v is the parent of u, we can write

$$n(u) = n(v) + \begin{cases} -\frac{h}{2^i} & \text{if } u \text{ is the left child of } v \\ +\frac{h}{2^i} & \text{otherwise} \end{cases} \tag{5}$$

Notice that if u is the left child of v we have $b_i = 0$ and $d_i = -\frac{h}{2^i}$; otherwise $b_i = 1$ and $d_i = \frac{h}{2^i}$. This observation, along with (4) and (5) combined, allows us to write

$$n(u) = h + \sum_{j=1}^{i-1} d_j + d_i = h + \sum_{j=1}^{i} d_j$$

completing the proof of the lemma.

As an example, consider again node u in Figure 2. Applying Lemma 2, one gets $n(u) = 4 - \frac{4}{2^1} + \frac{4}{2^2} = 3$.

With these technicalities out of the way, we now return to the corona training protocol. In our setting, the leaves of T represent the k coronas, while the preorder and inorder numbers of the nodes in T' are related, respectively, to the time slots in the training protocol and to the transmission ranges used by the sink. The goal of the training protocol is that all the sensors belonging to any corona c have to learn the $\log k$ bits, $b_1, b_2, \ldots, b_{\log k}$, which are the binary representation of their corona number minus one.

The corona training protocol consists of two phases: a first centralized phase, followed by a second distributed phase. The first phase is sink-driven and lasts for $2h - 1$ time slots. During this phase, the sensors learn the leftmost $\log h + 1$ bits, while the remaining bits will be learned in the second phase. At each time slot of the first phase, the sink transmits with a suitable power level and some sensors are awake to learn one more bit. The procedures performed by the sink node and the awake sensors in the centralized phase are described as follows.

Referring again to the T' tree, consider a generic time slot s_z, with $1 \leq z \leq 2h - 1$. Let u be the vertex of T' such that its preorder number $p(u) = z$, and let S_u be the subtree of T rooted at u (see e.g. Figure 2). At time slot s_z, the sink node transmits with a power level equal to $r_{\frac{k}{2h} n(u)}$, where $n(u)$ is the inorder number of node u, and the awake sensors are those belonging to the coronas which are the leaves of S_u. Although all the sensors in the coronas $1, \ldots, \frac{k}{2h} n(u)$ can hear the sink transmission, only those awake will learn one more bit. Precisely, the awake sensors that hear the sink transmission get $b_{i+1} = 0$, while the awake sensors that do not hear anything get $b_{i+1} = 1$, where i is the depth of node u in T'. It is worthy to note that, at time slot s_1, u is the root of T' and thus all the sensors are awake. As soon as a sensor has learned b_{i+1}, with $i \leq \log h$, it can easily compute the value c_{i+1} given in Lemma 1, and hence derive $p(u) + c_{i+1}$. If $p(u) + c_{i+1} = z + 1$ (i.e., $b_{i+1} = 0$), the sensor remains awake; otherwise, it goes to sleep. If $i < \log h$, then the sensor will wake up again at time slot $s_{p(u)+c_{i+1}}$. If $i = \log h$, let γ be the integer represented by the $\log h + 1$ bits learned so far by the sensor, namely $\gamma = \sum_{j=1}^{\log h+1} b_j 2^{\log h+1-j}$, then the sensor will wake up again at time slot $s_{2h+2\gamma}$.

In order to verify the correctness of the first phase of the corona training protocol, the following lemma is useful.

Lemma 3. *Let u be any node of T', with depth $i > 0$, and let v be the parent of u. The subtree S_u rooted at u contains $|n(v) - n(u)|\frac{k}{h}$ leaves, whose indices are:*

$$
\begin{cases}
(2n(u) - n(v)) \frac{k}{2h} + 1, \ \ldots, \ n(v)\frac{k}{2h} & \text{if } u \text{ is the left child of } v \\[2mm]
n(v)\frac{k}{2h} + 1, \ \ldots, \ (2n(u) - n(v)) \frac{k}{2h} & \text{if } u \text{ is the right child of } v
\end{cases}
$$

Proof. It follows immediately by the definition of the inorder number $n(u)$ and by the fact that any subtree S_u, with root at depth $i = \log h$, has $\frac{k}{h}$ leaves of T.

As an example, refer again to Figure 2, where the labels outside the nodes of T' give their preorder numbers, while those inside give their inorder numbers. Consider the node u having $p(u) = 4$ and $n(u) = 3$. The subtree S_u contains 4 leaves, indexed $5, 6, 7, 8$. Indeed, u is a right child, its parent v has $n(v) = 2$,

$|n(v)-n(u)|\frac{k}{h} = |2-3|\frac{16}{4} = 4$, $n(v)\frac{k}{2h}+1 = 2\frac{16}{8}+1 = 5$, and $(2n(u)-n(v))\frac{k}{2h} = (6-2)\frac{16}{8} = 8$.

Theorem 1. *Consider a time slot s_z, with $1 \leq z \leq 2h - 1$. At time s_z, all the sensors belonging to any corona c, with $1 \leq c \leq k$, have learned bits $b_1, b_2, \ldots, b_{i+1}$, where i is the depth of the deepest node u on the unique path from the root to leaf c such that $p(u) \leq z$.*

Proof. The proof is by induction on z. As the basis, observe that for $z = 1$, the root of T' is the only node with $p(u) \leq 1$. Observe that the depth of the root u is 0, all the sensors are awake, and the sink has transmitted with a power level $r_{n(u)\frac{k}{2h}} = r_{h\frac{k}{2h}} = r_{\frac{k}{2}}$. Therefore, all the sensors in any corona c have learned bit b_1, namely those in the first $\frac{k}{2}$ coronas have learned 0, and the others have learned 1.

For the inductive step, assume the statement true for $z - 1$. At time slot s_z, the only sensors awake are those belonging to the coronas which are the leaves of the subtree S_u, rooted at the node u such that $p(u) = z$. All the sensors in the other coronas are sleeping. Indeed, this is correct since the deepest node on the unique path from the root has not changed, and therefore such sensors have to learn no bits during this time slot. To check that the sensors in S_u learn the right bit, consider the node v such that $p(v) = z - 1$, and let w be the lowest common ancestor between u and v. Let ℓ be the depth of w. By inductive hypothesis, since $p(w) < p(u)$, all the sensors in S_u already know bits $b_1, \ldots, b_{\ell+1}$. At time s_z, the sink transmits with power level $r_{n(u)\frac{k}{2h}}$. Two cases may arise.

Case 1. Node u is the left child of w, that is $w = v$. By Lemma 3, the index of the middle corona among the leaves of S_u is $(2n(u) - n(v))\frac{k}{2h} + (n(v) - n(u))\frac{k}{2h} = n(u)\frac{k}{2h}$. Therefore, the sensors in the coronas $(2n(u) - n(v))\frac{k}{2h} + 1, \ldots, n(u)\frac{k}{2h}$ learn $b_{\ell+2} = 0$, while those in $n(u)\frac{k}{2h} + 1, \ldots, n(v)\frac{k}{2h}$ learn $b_{\ell+2} = 1$. Since the depth of u is $\ell + 1$, the statement is proved.

Case 2. Node u is the right child of w, and hence $w \neq v$. By Lemma 3, the index of the middle corona among the leaves of S_u is $n(w)\frac{k}{2h} + (n(u) - n(w))\frac{k}{2h} = n(u)\frac{k}{2h}$. Therefore, the sensors in the coronas $n(w)\frac{k}{2h} + 1, \ldots, n(u)\frac{k}{2h}$ learn $b_{\ell+2} = 0$, while those in $n(u)\frac{k}{2h} + 1, \ldots, (2n(u) - n(w))\frac{k}{2h}$ learn $b_{\ell+2} = 1$. Since the depth of u is $\ell + 1$, the statement is proved.

To illustrate Theorem 1, refer again to node u of Figure 2. Only the sensors in the leaves of S_u are awake in time slot $s_{p(u)} = s_4$, while the sink node transmits with a range of $r_{n(u)\frac{k}{2h}} = r_6$ since $\frac{k}{2h} = \frac{16}{8} = 2$ and $n(u) = 3$. The sensors in the leaves of S_u at a distance from the sink not exceeding r_6 will receive the signal, while the others will not. Since the depth of u is 2, the sensors in leaves 5 and 6 learn bit $b_3 = 0$, while those in leaves 7 and 8 learn bit $b_3 = 1$.

Corollary 1. *At time slot s_{2h-1}, the first phase of the corona training protocol is completed, and the sensors belonging to any corona c, with $1 \leq c \leq k$, have learned the leftmost $\log h + 1$ bits, $b_1, \ldots, b_{\log h+1}$, of the binary representation of their corona number minus one.*

Consider now the second phase of the corona training protocol, which starts at time slot s_{2h}. During such a phase, all the sensors have to learn the remaining $\log k - \log h - 1$ bits, $b_{\log h + 2}, \ldots, b_{\log k}$. Observe that there are $2h$ groups, each of $\frac{k}{2h}$ consecutive coronas which have learned the same $\log h + 1$ bits. Within each group, the sensors that belong to the first and last corona can become aware of their position by listening to the sink. Subsequently, in a distributed way, the sensors that have already known their position can inform those in the next corona to properly set their remaining bits.

The algorithm for the second phase is detailed as follows. Consider a generic group γ consisting of coronas $\gamma \frac{k}{2h} + 1, \ldots, (\gamma + 1)\frac{k}{2h}$, with $0 \leq \gamma \leq 2h - 1$. At the beginning of the second phase, all the sensors in such a group know $\gamma = \sum_{j=1}^{\log h + 1} b_j 2^{\log h + 1 - j}$, and wake up at time slot $s_{2h + 2\gamma}$. At time slot $s_{2h + 2\gamma}$, the sink transmits with a power level of $r_{(\gamma + 1)\frac{k}{2h} - 1}$. The sensors that do not hear it set every bit $b_{\log h + 2}, \ldots, b_{\log k}$ to 1 and go to sleep. At time slot $s_{2h + 2\gamma + 1}$, the sink transmits with a power level of $r_{\gamma \frac{k}{2h} + 1}$. The sensors that hear it set $b_{\log h + 2}, \ldots, b_{\log k}$ to 0, start the distributed computation by sending a message within their local transmission range, and then go to sleep. In a subsequent time slot s_t, an awake sensor that receives a message from another sensor computes $\delta = t - (2h + 2\gamma + 1)$, sets its bits $b_{\log h + 2}, \ldots, b_{\log k}$ to the binary representation of δ (with the most significant bit assigned to $b_{\log_{h+2}}$), and goes to sleep. Therefore, the following result easily holds.

Lemma 4. *All the sensors belonging to corona c, with $1 \leq c \leq k$, have learned the binary representation of $c - 1$ at time slot*

$$\begin{cases} s_{2h + 2\gamma + 1 + \delta} & \text{if } c = \gamma \frac{k}{2h} + \delta + 1 \\ s_{2h + 2\gamma} & \text{if } c = (\gamma + 1)\frac{k}{2h} \end{cases}$$

where $0 \leq \gamma \leq 2h - 1$ and $0 \leq \delta \leq \frac{k}{2h} - 1$.

The sensors in corona $k - 1$ are the last to learn all their bits. Since they belong to the $\left(\frac{k}{2h} - 1\right)$-th corona of group $\gamma = 2h - 1$, this happens at time $s_{(2h + 2(2h-1)+1) + \frac{k}{2h} - 1}$. Thus, the entire corona training protocol finishes at time $s_{6h - 2 + \frac{k}{2h}}$. Therefore, the overall time to accomplish the above corona training protocol is $O(h + \frac{k}{h})$. Such a time is minimized when $h = O(\sqrt{k})$, and in such a case it becomes $O(\sqrt{k})$, improving over the $O(k)$ time of the training protocol presented in [16].

It is also worth noting that only the sensor nodes that need to be awake in a given time slot will stay awake, the others will sleep minimizing the power expenditure. Yet another interesting feature of the training protocol is that individual sensors sleep for as many contiguous time slots as possible before waking up, thus avoiding repeated wake-sleep transitions that are known to waste energy. To see this, observe that the sensors remain awake $O(\log h)$ time during the first phase because they wake up just at the time slot when they have to learn one more bit. Moreover, the sensors are awake for $O(\frac{k}{h})$ consecutive time slots during the second phase. Therefore, the sensor awake time is $O(\log h + \frac{k}{h})$, which is

Table 1. Performance of the corona training protocol. The sensor awake time is optimal when $h = O\left(\frac{k}{\log k}\right)$, while the overall time is minimized when $h = O(\sqrt{k})$.

Performance Measure	Complexity	$h = O(\sqrt{k})$	$h = O\left(\frac{k}{\log k}\right)$
Overall Time	$O\left(h + \frac{k}{h}\right)$	$O(\sqrt{k})$	$O\left(\frac{k}{\log k}\right)$
Sensor Awake Time	$O\left(\log h + \frac{k}{h}\right)$	$O(\sqrt{k})$	$O(\log k)$
♯ Wake/Sleep Transitions	$O(\log h)$	$O(\log k)$	$O\left(\log\left(\frac{k}{\log k}\right)\right)$

minimized when $h = O(\frac{k}{\log k})$. In such a case the sensor awake time is $O(\log k)$ which is optimal since every sensor has to learn $\log k$ bits and it cannot learn more than one bit at a time. In addition, the number of wake-sleep transitions is $O(\log h)$. Precisely, referring again to Figure 2, one notes that a wake-sleep transition occurs every time a node u on the path of T' from the root to a generic corona is a right child of its parent. This is because the preorder numbers of u and its parent are not consecutive. Thus, the worst case arises for the sensors in the coronas of group $2h - 1$, which go through exactly $\log h + 1$ transitions during the first phase, plus one more transition during the second phase. The performance achieved by the corona training protocol for the above mentioned measures are summarized in Table 1.

5 Concluding Remarks

In this work a new training protocol has been proposed which outperforms that originally presented in [16] in terms of the overall time for training, lowering it from a linear to a square-root function of the size of the coordinate system used for location awareness. However, a number of questions still remain open. In particular, a good idea for further work should be that of comparing the performance of the protocol proposed in the present paper with that devised in [18]. Indeed, while the training protocol of Section 4 requires individual sensors to be awake for a short time but consumes energy in both the synchronization between the sensors and the sink and the toggling between sleep and wake periods, the asynchronous protocol proposed in [18] may force sensors to be awake for longer periods but avoids frequent transitions from sleep to wake periods.

References

1. F. Akyildiz, W. Su, Y. Sankarasubramanian, and E. Cayirci. Wireless sensor networks: a survey. *Computer Networks*, 38(4):393–422, 2002.
2. S. Bandyopadhyay and E. Coyle. An efficient hierarchical clustering algorithm for wireless sensor networks. In *Proc. IEEE INFOCOM 2003*, San Francisco, CA, April 2003.
3. D. Culler, D. Estrin, and M. Srivastava. Overview of sensor networks. *IEEE Computer*, 37(8):41–49, 2004.

4. D. Culler and W. Hong. Wireless sensor networks. *Communications of the ACM*, 47(6):30–33, 2004.
5. C.C. Enz, A. El-Hoiydi, J.-D. Decotignie, and V. Peiris. WiseNET: a ultralow power wireless sensor network solution. *IEEE Computer*, 37(8):62–69, 2004.
6. S. Ghiasi, A. Srivastava, X. Yang, and M. Sarrafzadeh. Optimal energy-aware clustering in sensor networks. *Sensors*, 2:258–269, 2002.
7. J. Hill, M. Horton, R. Kling, and L. Krishnamurthy. The platforms enabling wireless sensor networks. *Communications of the ACM*, 47(6):41–46, 2004.
8. K. Langendoen and N. Reijers. Distributed localization algorithm. In *Embedded Systems Handbook*, R. Zurawski (Editor), CRC Press, Boca Raton, FL, 2004.
9. K. Martinez, J.K. Hart, and R. Ong. Sensor network applications. *IEEE Computer*, 37(8):50–56, 2004.
10. D. Nicolescu. Positioning in ad-hoc sensor networks. *IEEE Network*, 18(4):24–29, 2004.
11. S. Olariu, A. Waada, L. Wilson, and M. Eltoweissy. Wireless sensor networks leveraging the virtual infrastructure. *IEEE Network*, 18(4):51–56, 2004.
12. P. Saffo. Sensors, the next wave of innovation. *Communications of the ACM*, 40(2):93–97, 1997.
13. M. Sichitiu and C. Veerarithiphan. Simple accurate synchronization for wireless sensor networks. In *Proc. WCNC'03*, 2003.
14. F. Sivrukaya and B. Yener. Time synchronization in sensor networks: a survey. *IEEE Network*, 18(4):45–50, 2004.
15. K. Sohrabi, J. Gao, V. Ailawadhi, and G. Pottie. Protocols for self-organization of a wireless sensor network. *IEEE Personal Communications*, 16–27, October 2000.
16. A. Waada, S. Olariu, L. Wilson, M. Eltoweissy, and K. Jones. Training a wireless sensor network. *Mobile Networks and Applications*, 10(1):151–168, 2005.
17. B. Warneke, M. Last, B. Leibowitz, and K. Pister. SmartDust: communicating with a cubic-millimeter computer. *IEEE Computer*, 34(1):44–51, 2001.
18. Q. Xu, R. Ishak, S. Olariu, and S. Salleh. On asynchronous training in sensor networks. In *3rd Intl. Conf. on Advances in Mobile Multimedia*, K.Lumpur, September 2005.
19. V.V. Zhirnov, and D.J.C. Herr. New frontiers: self-assembly and nano-electronics. *IEEE Computer*, 34(1):34–43, 2001.

On the Complexity of Minimizing Interference in Ad-Hoc and Sensor Networks[*]

Davide Bilò[1] and Guido Proietti[1,2]

[1] Dipartimento di Informatica, Università di L'Aquila, Italy
[2] Istituto di Analisi dei Sistemi ed Informatica, CNR, Roma, Italy
{davide.bilo, proietti}@di.univaq.it

Abstract. One of the most critical factors for lifetime and operability of ad-hoc and sensor networks is the limited amount of available energy. To this respect, minimizing the interference in the network has certainly a positive effect, since of the reduced number of conflicting transmissions. However, quite surprisingly, only few theoretical results are known about the possibility to maintain the interference low while at the same time guaranteeing certain network connectivity properties. In this paper, we give a contribution in this direction, and we study several network interference measures with respect to the symmetric connectivity, the strong connectivity, and the broadcast connectivity predicate.

In particular, we show that the probably most prominent interference problem, namely that of minimizing the *maximum* interference experienced by any node in the network, is hard to approximate within better than a logarithmic factor, unless NP admits slightly superpolynomial time algorithms, for any of the above connectivity predicates. On a positive side, we show that any approximation algorithm for the problem of minimizing the total range assigned to the nodes in order to guarantee any of the above connectivity predicates, can be transformed, by maintaining the approximation ratio, into an approximation algorithm for the problem of minimizing the *total* interference experienced by all the nodes in the network.

Keywords: Ad-Hoc Networks, Sensor Networks, Interference, Range Assignment.

1 Introduction

During the last decade wireless networks have received significant attention because of the recent drop in equipment prices. In particular, ad-hoc and sensor networks have emerged due to their great potential in many application scenarios such as battlefield, disaster relief, emergency, monitoring and surveillance, and data gathering.

[*] Work partially supported by the Research Project GRID.IT, funded by the Italian Ministry of Education, University and Research and by the European Union under IST FET Integrated Project 015964 AEOLUS and COST Action 293 GRAAL.

Ad-hoc and sensor networks consist of autonomous battery-operated *devices* that can communicate each other by means of radio signals and without relaying on fixed communication infrastructures. Because of the limited amount of energy, the communication reflects the *multi-hop* model principles, i.e., the two parties communicate directly if they are close enough, otherwise they rely to intermediate devices. Moreover, devices are equipped with specific components allowing them to adjust their transmission power. Clearly, signal intensity, transmission power, and energy consumption are quantities proportionally related.

In the general signal transmission model each device transmits a message at a certain intensity which may attenuate during propagation because of local climatic condition, interference from other transmitting devices, presence of obstacles, etc. Since receivers can correctly read a message if the intensity at whom it is perceived is not lower than some threshold value, namely the *transmission quality parameter*, then the transmission power of a device affects its *transmission range*, i.e., the set of devices it can directly send messages to.

Since energy is one of the most limiting factors for lifetime and operability of ad-hoc and sensor networks, then many research activities have focused on the design of algorithms for minimizing the overall range power assigned to the stations [2,5,3,9,13,17]. However, another approach to achieve energy conservation is that of minimizing interference. Indeed, lowering interference causes a reduction on the number of collisions and thus a saving on the energy spent for packets retransmission on the *Media Access Layer*. In the past, most of the previous works implicitly tried to contain interference by means of *topology control* approaches, e.g., by constructing network structures having desirable properties such as sparseness or low node-degree [8,10,11,12,16]. Unfortunately, as observed in [1,18], these two features are not sufficient to contain interference. Hence, researchers are currently addressing the problem of containing interference directly.

In the literature, different models for interference have been proposed by the research community [1,7,14,18]. In this paper, we focus on the two most relevant models among the ones measuring interference at devices. They are the following:

Interference of the Sender (SI model): is the interference caused by a device s transmitting at power p. Its measure is equal to the cardinality of the set of devices s can directly send messages to;

Interference at the Receiver (RI model): is the interference experienced by a device v, and it is measured as the cardinality of the set of devices which can directly send messages to v.

Of the two, the RI model seems to better reflects the intuition of the real world. Indeed, as argued in [18], (i) message collisions occurs at receivers, and (ii) the RI model is more robust than the SI one in terms of measure increase due to the arrival of additional devices in the network.

For this reason, in this paper we mainly focus on the RI model. More precisely, we propose different reasonable optimization problems for the above interference models, and we prove approximability and/or non-approximability results for interesting connectivity predicates such as symmetric connectivity, strong connectivity and broadcast connectivity.

Organization of the paper. The remainder of the paper is organized as follows: in Section 2 we give the basic definitions we will use throughout the rest of the paper; in Section 3 we define some interference problems and the corresponding known results in the literature, while in Section 4 we present our results.

2   Preliminaries

Let V be a set of n *devices* and let $r : V \times V \longrightarrow \mathbb{R}^+ \cup \{+\infty\}$ be the *potential transmission range function*. More precisely, $\forall u, v \in V$, $r(u, v)$ represents the power level at whom u must transmits in order to send a message to v in one-hop. Let $\mathcal{R} : V \longrightarrow \mathbb{R}^+$ be the *range assignment*, i.e., the transmission power level of the devices. For a transmission range assignment $\mathcal{R} : V \longrightarrow \mathbb{R}^+$:

- the *asymmetric network* $D_\mathcal{R}$ *induced from* $\mathcal{R}$: is a network that can be modelled with a digraph $D_\mathcal{R} = (V, A)$, where A contains the directed edge (u, v) iff $\mathcal{R}(u) \geq r(u, v)$;
- the *symmetric network* $G_\mathcal{R}$ *induced from* $\mathcal{R}$: is a network that can be modelled with a graph $G_\mathcal{R} = (V, E)$, where E contains the undirected edge $\{u, v\}$ iff $\mathcal{R}(u) \geq r(u, v) \wedge \mathcal{R}(v) \geq r(v, u)$.

We say that a range assignment $\mathcal{R}$ *satisfies* a connectivity predicate π, if the asymmetric/symmetric network induced from $\mathcal{R}$ satisfies π. For any two range assignments $\mathcal{R}, \mathcal{R}'$, we write $\mathcal{R} \preceq \mathcal{R}'$ if $\forall v \in V, \mathcal{R}(v) \leq \mathcal{R}'(v)$. Moreover we say that $\mathcal{R}$ is *topologically minimal* w.r.t. connectivity predicate π if there is no $\mathcal{R}' \neq \mathcal{R}$ satisfying π and such that $\mathcal{R}' \preceq \mathcal{R}$. Finally, given a range assignment $\mathcal{R}$, and a device $v \in V$, we define:

- the *set of devices* $I_{\mathcal{R},\mathsf{out}}(v)$ *interfered by the sender* v (SI model):

$$I_{\mathcal{R},\mathsf{out}}(v) = \{u \in V \setminus \{v\} \mid r(v, u) \leq \mathcal{R}(v)\};$$

- the *set of devices* $I_{\mathcal{R},\mathsf{in}}(v)$ *interfering at the receiver* v (RI model):

$$I_{\mathcal{R},\mathsf{in}}(v) = \{u \in V \setminus \{v\} \mid r(u, v) \leq \mathcal{R}(u)\}.$$

3   Definition of the Interference Problems

In this section we define some interference problems for ad-hoc and sensor networks. The input is a set V of n devices, a potential transmission range function $r : V \times V \longrightarrow \mathbb{R}^+ \cup \{+\infty\}$ and a connectivity predicate π. The output is a range assignment $\mathcal{R} : V \longrightarrow \mathbb{R}^+$, so that $\mathcal{R}$ satisfies π. The objective is to minimize a function depending on the problem. Some interesting objective functions are the following:

- **The MinMax SI Problem** (MINMAXSIP):

$$I_{\mathsf{out}}(\mathcal{R}) = \max_{v \in V} |I_{\mathcal{R},\mathsf{out}}(v)|;$$

- **The Minimum Total Interference Problem (MTIP):**

$$I(\mathcal{R}) = \sum_{v \in V} |I_{\mathcal{R},\text{in}}(v)| = \sum_{v \in V} |I_{\mathcal{R},\text{out}}(v)|;$$

- **The Minimum RI Problem for a Given Device (MRIPGD):** takes $\nu \in V$ as additional input. The objective function is:

$$|I_{\mathcal{R},\text{in}}(\nu)|;$$

- **The MinMax RI Problem (MINMAXRIP):**

$$I_{\text{in}}(\mathcal{R}) = \max_{v \in V} |I_{\mathcal{R},\text{in}}(v)|.$$

Known results and related work. For the MINMAXSIP, in [1] the authors have designed polynomial time algorithms computing an optimal solution for the connectivity, strong connectivity and spanner predicates.

Concerning MTIP for connectivity, the results contained in [15] can be directly used to show that the problem can be approximated within $\mathcal{O}(\log n)$, and cannot be approximated within a factor better than $c \ln n$, for some constant $c > 0$, unless $\mathsf{NP} \subseteq \mathsf{DTIME}(n^{\mathcal{O}(\log \log n)})$, even if r is a *distance function*.[1] Moreover, in the same paper, the authors extended their approximability result to any connectivity predicate that can be formulated as a *0–1 proper function* [4].

Finally, for the MINMAXRIP for connectivity, in [18] it is presented an algorithm that achieves a $\mathcal{O}(\sqrt[4]{\Delta})$-approximation, where $\Delta = \max_{v \in V} |\{u \in V \setminus \{v\} \mid r(v,u) \neq +\infty\}|$ for the *highway model* (i.e., 1-dimensional Euclidean space) in *unit disk graphs*.[2]

4   Our Results

In this section, we present some results for all the optimization problems introduced in the previous section. In the remaining of the paper, we will focus on the most interesting communication topologies, that is *connectivity* (i.e., $G_{\mathcal{R}}$ is connected), *strong connectivity* (i.e, $D_{\mathcal{R}}$ is strong connected), and *broadcast* (i.e, $D_{\mathcal{R}}$ contains a directed path from a given device s to every $v \in V$).

All the non-approximability results we prove for the MRIPGD, and the MINMAXRIP are obtained by reductions from the SET COVER PROBLEM (SCP). An

---

[1] We recall that r is a *distance function* if the following properties hold:

- $r(v,v) = 0, \forall v \in V$;
- $r(u,v) = r(v,u), \forall u,v \in V$ (**symmetry**);
- $r(u,v) \leq r(u,w) + r(w,v), \forall u,v,w \in V$ (**triangle inequality**).

[2] In a *unit disk graph*, nodes are spread on a k-dimensional Euclidean space and $r(u,v)$ is equal to the Euclidean distance from u and v if it is not greater than 1, otherwise $r(u,v) = +\infty$.

instance $\mathcal{I} = \langle O, \mathcal{S} \rangle$ for the SCP consists of a set $O = \{o_1, \ldots, o_h\}$ of h objects, and a set $\mathcal{S} = \{S_1, \ldots, S_\ell\}$ of ℓ subsets of O. The objective is to find a minimum-size collection of subsets in $\mathcal{S}$ whose union is O. In [6] it is shown that SCP cannot be approximated within $(1 - o(1)) \ln h$, unless $\mathsf{NP} \subseteq \mathsf{DTIME}(h^{\mathcal{O}(\log \log h)})$ even for the case $\ell \leq h$.

Concerning approximability results, for the MINMAXSIP we design an optimal polynomial time algorithm for a large class of connectivity predicates, thus generalizing some of the results in [1]. For the MTIP and the MRIPGD, we reduce them to the *Minimum Total Range Assignment Problem* (MTRAP), a problem which asks for a range assignment $\mathcal{R}$ satisfying π and such that the following objective function

$$cost(\mathcal{R}) = \sum_{v \in V} \mathcal{R}(v)$$

is minimized. For the MTRAP, in Table 1 we summarize the known results proved in [2,5].

Table 1. Table of known results for MTRAP. By **SCH** we mean that the problem is as hard as the SCP, i.e., it cannot be approximated within a factor better $c \log n$, for some constant $c > 0$, unless $\mathsf{NP} \subseteq \mathsf{DTIME}(n^{\mathcal{O}(\log \log n)})$.

π	APPROXIMABILITY	NON-APPROXIMABILITY
CONNECTIVITY	$2 + 2\ln(n-1)$ [2]	**SCH** [2]
STRONG CONNECTIVITY	$3 + 2\ln(n-1)$ [2]	**SCH** [2]
BROADCAST	$2 + 2\ln(n-1)$ [2]	**SCH** [5]

4.1   Results for MINMAXSIP

We design an optimal algorithm OPT-MINMAXSIP for the MINMAXSIP and we show that such an algorithm runs in polynomial time for a class of connectivity predicates.

First, $\forall v \in V$ we define the set $P(v) = \{r(v, u) \mid u \in V \setminus \{v\}\}$. Next, $\forall p \in P(v)$ let $\phi(v, p)$ be the number of devices interfered by the sender v if its range assignment is p, i.e., $\phi(v, p) = |\{u \in V \setminus \{v\} \mid r(v, u) \leq p\}|$. Finally, let P be the multiset made up of the union of all $P(v)$'s. The pseudocode of algorithm OPT-MINMAXSIP is reported below:

Lemma 1. *Algorithm 1 returns an optimal solution.*

Proof. For the sake of contradiction, assume that Algorithm 1 does not return an optimal solution. Hence, there must be a feasible range assignment $\mathcal{R}'$ such that $I_{out}(\mathcal{R}') < I_{out}(\mathcal{R})$. As a consequence $\mathcal{R}' \preceq \mathcal{R}$. We have obtained a contradiction on the stop criterion of the while loop of the algorithm.   □

Definition 1. *A connectivity predicate π is said to be a P-connectivity predicate if the decision problem of determining whether a (di)graph contains a subgraph satisfying π is polynomial time solvable.*

Algorithm 1. OPT-MINMAXSIP

Input: a set of devices V, a potential range assignment $r : V \times V \longrightarrow \mathbb{R}^+$, and a connectivity predicate π.

Output: a range assignment $\mathcal{R}$ satisfying π.

  1: sort P in a non-decreasing way w.r.t. $\phi(\cdot, \cdot)$. Denote by $r(v_1, u_1), \ldots, r(v_{|P|}, u_{|P|})$
     the ordered sequence.
  2: **for all** $v \in V$ **do**
  3:    $\mathcal{R}(v) = 0$
  4: **end for**
  5: $i = 1$
  6: **while** $\mathcal{R}$ does not satisfy π **do**
  7:    $\mathcal{R}(v_i) = r(v_i, u_i)$
  8:    $i = i + 1$
  9: **end while**
10: **return** $\mathcal{R}$

Theorem 1. *If π is a P-connectivity predicate, then* MINMAXSIP *for π is polynomial time solvable.*

Proof. From Lemma 1 we can use algorithm OPT-MINMAXSIP for computing an optimal solution. Moreover its running time is polynomial time since π is a P-connectivity predicate and since i is upper bounded by $|P| = \mathcal{O}(n^2)$. $\square$

From Theorem 1, it follows that MINMAXSIP restricted to connectivity, strong connectivity, and broadcast is polynomial-time solvable. Moreover, other interesting P-connectivity predicates are: k-edge (resp., k-vertex) connectivity, spanners, and k in-branching (resp., k out-branching).

4.2 Results for MTIP

In this section we present a reduction from the MTIP to the MTRAP. Some of the results we obtain are similar to those proved in [15]. However, our proofs are simpler, and address the problem under a different perspective thus showing its relationship with the MTRAP.

Theorem 2. *Let π be a connectivity predicate. Any ρ-approximation algorithm for the MTRAP for π can be used to find a ρ-approximation for the MTIP for π.*

Proof. Let $\mathcal{I} = \langle V, r, \pi \rangle$ be an instance for MTIP. From $\mathcal{I}$ we build an instance $\mathcal{I}' = \langle V, r', \pi \rangle$ for MTRAP in the following way:

$$- r'(u, v) = \begin{cases} +\infty & \text{if } r(u, v) = +\infty; \\ |\{w \in V \setminus \{u\} \mid r(u, w) \leq r(u, v)\}| & \text{otherwise.} \end{cases}$$

Let $\mathcal{R}'$ be a feasible range assignment for $\mathcal{I}'$. We define a range assignment $\mathcal{R} = \psi(\mathcal{R}')$ for $\mathcal{I}$ from $\mathcal{R}'$ in the following way:

$$\mathcal{R}(v) = \max\{r(v,u) \mid u \in V \setminus \{v\}, r'(v,u) \leq \mathcal{R}'(v)\}.$$

Notice that ψ is a bijection. It is not hard to see that $\mathcal{R}'$ and $\psi(\mathcal{R}')$ induces the same asymmetric/symmetric network. Moreover

$$\mathsf{cost}(\mathcal{R}') = \sum_{v \in V} \mathcal{R}'(v) = \sum_{v \in V} |I_{\mathcal{R},\mathsf{out}}(v)| = I(\mathcal{R}). \qquad \square$$

From the above theorem, we can immediately infer for MTIP the approximability results contained in Table 2.

Table 2. Table of results for MTIP. By **SCH** we mean that the problem is as hard as the SCP, i.e., it cannot be approximated within a factor better than $c\log n$, for some constant $c > 0$, unless NP $\subseteq$ DTIME$(n^{\mathcal{O}(\log\log n)})$. The question mark means that the corresponding entry of the table is an open problem.

π	APPROXIMABILITY	NON-APPROXIMABILITY
CONNECTIVITY	$2 + 2\ln(n-1)$	**SCH** [15]
STRONG CONNECTIVITY	$3 + 2\ln(n-1)$	?
BROADCAST	$2 + 2\ln(n-1)$	?

Remark 1. We want to point out that, if the instance of the MTRAP generated in Theorem 2 from the instance of the MTIP is symmetric (i.e., $r(u,v) = r(v,u), \forall u, v, \in V$), then we can compute a solution for the MTIP with a better performance guarantee since better approximation algorithms for the symmetric instances of MTRAP are known (see [5,3,17]).

4.3   Results for MRIPGD

Theorem 3. *Let π be a connectivity predicate. Any ρ-approximate algorithm for MTRAP for π can be used to find a ρ-approximation for the MRIPGD for π.*

Proof. Let $\mathcal{I} = \langle V, \nu, r, \pi \rangle$ be an instance for MRIPGD. From $\mathcal{I}$ we build an instance $\mathcal{I}' = \langle V, r', \pi \rangle$ for MTRAP in the following way:

$$- r'(u,u') = \begin{cases} +\infty & \text{if } r(u,u') = +\infty; \\ 1 & \text{if } u \neq \nu \text{ and } r(u,\nu) \leq r(u,u'); \\ 0 & \text{otherwise.} \end{cases}$$

Let $\mathcal{R}'$ be a feasible range assignment for $\mathcal{I}'$. We define a range assignment $\mathcal{R} = \psi(\mathcal{R}')$ for $\mathcal{I}$ from $\mathcal{R}'$ in the following way:

$$\mathcal{R}(v) = \max\{r(v,u) \mid u \in V \setminus \{v\}, r'(v,u) \leq \mathcal{R}'(v)\}$$

Notice that ψ is a bijection. It is not hard to see that $\mathcal{R}'$ and $\psi(\mathcal{R}')$ induces the same asymmetric/symmetric network. Moreover

$$\text{cost}(\mathcal{R}) = \sum_{v \in V} \mathcal{R}'(v) = |I_{\mathcal{R},\text{in}}(\nu)|. \qquad \square$$

Remark 2. As for the case of the MTIP (see Remark 1), if the instance of the MTRAP generated is symmetric when we apply the reduction defined in the proof of Theorem 3, then we can compute a solution for the MRIPGD with a better performance guarantee.

Proposition 1. MRIPGD *for connectivity is not approximable within* $(1 - o(1)) \ln \frac{n-2}{2}$, *unless* $\mathsf{NP} \subseteq \mathsf{DTIME}(n^{\mathcal{O}(\log \log n)})$, *even if* r *is a distance.*

Proof. Let $\mathcal{I} = \langle O, \mathcal{S} \rangle$ be an instance for the SCP with $\ell \leq h$. From $\mathcal{I}$ we build in polynomial time an instance $\mathcal{I}'$ for MRIPGD for connectivity as follows.

The set of devices is $V = \{\nu, u, v_1, \ldots, v_h, s_1, \ldots, s_\ell\}$, where v_j is the representative of object o_j, while s_j is the representative of set S_j. The device to whom we measure the interference is ν.

The potential transmission range function is defined as follows:

- $r(u, s_j) = 2, \forall S_j \in \mathcal{S}$;
- $r(s_j, \nu) = 3, \forall S_j \in \mathcal{S}$;
- $r(s_j, v_k) = 3, \forall S_j \in \mathcal{S}, \forall o_k \in S_j$;
- $r(v_k, \nu) = 5, \forall o_k \in O$.

The potential transmission range function for all the other pairs of devices is equal to the *metric closure*.[3]

Let $\text{SOL}(\mathcal{I})$ be a solution for the instance $\mathcal{I}$ of SCP. From $\text{SOL}(\mathcal{I})$ we build a solution for the instance of MRIPGD for connectivity as follows:

- $\mathcal{R}(\nu) = 3$;
- $\mathcal{R}(u) = 2$;
- $\mathcal{R}(v_j) = 3, \forall j = 1, \ldots, h$;
- $\mathcal{R}(s_j) = 3, \forall S_j \in \text{SOL}(\mathcal{I})$;
- $\mathcal{R}(s_j) = 2, \forall S_j \notin \text{SOL}(\mathcal{I})$.

Since $\text{SOL}(\mathcal{I})$ is a solution for the instance $\mathcal{I}$ of the SCP, then it follows that $G_{\mathcal{R}}$ is connected. Moreover the interference experienced by ν is $|I_{\mathcal{R},\text{in}}(\nu)| = |\text{SOL}(\mathcal{I})|$.

Now in order to prove the lemma, it remains to show that any feasible range assignment $\mathcal{R}$ for MRIPGD for connectivity can be used for building a solution for $\mathcal{I}$ of SCP with size not exceeding $|I_{\mathcal{R},\text{in}}(\nu)|$. First, starting from $\mathcal{R}$ we define a new feasible range assignment $\mathcal{R}'$ such that $|I_{\mathcal{R}',\text{in}}(\nu)| \leq |I_{\mathcal{R},\text{in}}(\nu)|$. We start by assigning $\mathcal{R}'(v) = \mathcal{R}(v)$ to every device v. For each $v_k \in I_{\mathcal{R},\text{in}}(\nu)$, we choose a set

---

[3] We mean that $\forall u, v \in V, r(u, v)$ is equal to the total cost of the shortest path from u to v.

$S_j \in \mathcal{S}$ containing object o_k and set $\mathcal{R}'(s_j) = \mathcal{R}'(v_k) = 3$. Finally, we set $\mathcal{R}'(u)$ to 2. Notice that $G_{\mathcal{R}'}$ is still connected and the number of stations interfering with ν is not greater than $I_{\mathcal{R},\text{in}}(\nu)$.

We define a solution for instance $\mathcal{I}$ of the SCP as the set of all S_j such that $s_j \in I_{\mathcal{R}',\text{in}}(\nu)$. The claim follows from the inapproximability of the SCP [6].   □

Proposition 2. MRIPGD *for strong connectivity is not approximable within* $(1 - o(1)) \ln \frac{n-2}{2}$, *unless* NP $\subseteq$ DTIME$(n^{\mathcal{O}(\log \log n)})$, *even if r is symmetric.*

Proof. Let $\mathcal{I} = \langle O, \mathcal{S} \rangle$ be an instance for the SCP with $\ell \leq h$. From $\mathcal{I}$ we build in polynomial time an instance $\mathcal{I}'$ for MRIPGD for strong connectivity as it was done in Lemma 1 except for the potential transmission range function which is defined as follows:

- $r(u, s_j) = 2, \forall S_j \in \mathcal{S}$;
- $r(s_j, \nu) = 3, \forall S_j \in \mathcal{S}$;
- $r(s_j, v_k) = 3, \forall S_j \in \mathcal{S}, \forall o_k \in S_j$.

The potential transmission range function of all other pairs of devices is set to $+\infty$. Let $\text{SOL}(\mathcal{I})$ be a solution for the instance $\mathcal{I}$ of SCP. From $\text{SOL}(\mathcal{I})$ we build a solution $\mathcal{R}$ for the instance of MRIPGD for strong connectivity as it was done in Lemma 1. It is easy to see that $D_{\mathcal{R}}$ is strong connected. Moreover $|I_{\mathcal{R},\text{in}}(\nu)| = |\text{SOL}(\mathcal{I})|$.

To conclude the proof, consider a feasible range assignment $\mathcal{R}$ such that $D_{\mathcal{R}}$ is strong connected. We define a solution for instance $\mathcal{I}$ of the SCP as the set of all S_j such that $s_j \in I_{\mathcal{R},\text{in}}(\nu)$. Now the claim follows from the inapproximability of the SCP [6].   □

Proposition 3. MRIPGD *for broadcast is not approximable within* $(1 - o(1)) \ln \frac{n-2}{2}$, *unless* NP $\subseteq$ DTIME$(n^{\mathcal{O}(\log \log n)})$, *even if r is symmetric.*

Proof. Let $\mathcal{I} = \langle O, \mathcal{S} \rangle$ be an instance for SCP with $\ell \leq h$. From $\mathcal{I}$ we build in polynomial time an instance $\mathcal{I}'$ for MRIPGD for broadcast as it was done in Lemma 2. The root is u. Let $\text{SOL}(\mathcal{I})$ be a solution for the instance $\mathcal{I}$ of SCP. The range assignment $\mathcal{R}$ is defined as follows:

- $\mathcal{R}(u) = 2$;
- $\mathcal{R}(s_j) = 3, \forall S_j \in \text{SOL}(\mathcal{I})$;

It is easy to see that $D_{\mathcal{R}}$ is a spanning arborescence rooted at u. Moreover the interference experienced by ν is $|I_{\mathcal{R},\text{in}}(\nu)| = |\text{SOL}(\mathcal{I})|$. To conclude the proof, let $\mathcal{R}$ be a range assignment such that $D_{\mathcal{R}}$ contains a spanning arborescence rooted at u. The solution for SCP is given by the set of all S_j such that $s_j \in I_{\mathcal{R},\text{in}}(\nu)$. The claim follows from the inapproximability of SCP [6].   □

Table 3. Table of results for MRIPGD

π	APPROXIMABILITY	NON-APPROXIMABILITY
CONNECTIVITY	$2 + 2\ln(n-1)$	$(1 - o(1))\ln\frac{n-2}{2}$ unless $\mathsf{NP} \subseteq \mathsf{DTIME}(n^{\mathcal{O}(\log\log n)})$
STRONG CONNECTIVITY	$3 + 2\ln(n-1)$	$(1 - o(1))\ln\frac{n-2}{2}$ unless $\mathsf{NP} \subseteq \mathsf{DTIME}(n^{\mathcal{O}(\log\log n)})$
BROADCAST	$2 + 2\ln(n-1)$	$(1 - o(1))\ln\frac{n-2}{2}$ unless $\mathsf{NP} \subseteq \mathsf{DTIME}(n^{\mathcal{O}(\log\log n)})$

Table 3 summarizes the results we obtained for MRIPGD.

4.4 Results for MINMAXRIP

Proposition 4. MINMAXRIP *for connectivity is not approximable within a factor better than* $c\ln n$, *for some constant* $c > 0$, *unless* $\mathsf{NP} \subseteq \mathsf{DTIME}(n^{\mathcal{O}(\log\log n)})$, *even if* r *is a distance function.*

Proof. Let $\mathcal{I} = \langle O, \mathcal{S} \rangle$ be an instance for SCP with $\ell \leq h$ and let $t = h + 2$. From $\mathcal{I}$ we build in polynomial time an instance $\mathcal{I}'$ for MINMAXRIP for connectivity as follows. The set of devices is $V = \{\nu\} \cup \bigcup_{i=1}^{t} V_i$, where $V_i = \{u_i, v_i^1, \ldots, v_i^h, s_i^1, \ldots, s_i^\ell\}$. For each $i = 1, \ldots, t$, s_i^j is the representative for set S_j, while v_i^k is the representative for object o_k. The potential transmission range function is defined as follows:

- $r(u_i, s_i^j) = 2, \forall i = 1, \ldots, t, \forall S_j \in \mathcal{S}$;
- $r(s_i^j, \nu) = 3, \forall i = 1, \ldots, t, \forall S_j \in \mathcal{S}$;
- $r(s_i^j, v_i^z) = 3, \forall i = 1, \ldots, t, \forall S_j \in \mathcal{S}, \forall o_z \in S_j$;
- $r(v_i^j, \nu) = 5, \forall i = 1, \ldots, t, \forall o_j \in O$.

The potential transmission range function for all other pairs of devices is equal to the metric closure.

Before proving the claim, let us have a look at the lower bounds for the interference experienced by the devices. Consider any feasible range assignment $\mathcal{R}$. $\forall i \in \{1, \ldots, t\}$, it is not hard to see that:

- $\ell \leq |I_{\mathcal{R},\mathsf{in}}(u_i)|$;
- $|S_j| + 2 \leq |I_{\mathcal{R},\mathsf{in}}(s_i^j)|, \forall j = 1, \ldots, \ell$;
- $1 \leq |I_{\mathcal{R},\mathsf{in}}(v_i^k)|, \forall k = 1, \ldots, h$;
- $t \leq |I_{\mathcal{R},\mathsf{in}}(\nu)|$.

Hence $I_{\mathsf{in}}(\mathcal{R}) \geq t$. Let $\mathrm{SOL}(\mathcal{I})$ be a solution for the instance $\mathcal{I}$ of the SCP. From $\mathrm{SOL}(\mathcal{I})$ we build a solution for the instance of MINMAXRIP as follows:

- $\mathcal{R}(\nu) = 3$;
- $\mathcal{R}(u_i) = 2, \forall i = 1, \ldots, t$;
- $\mathcal{R}(v_i^k) = 3, \forall i = 1, \ldots, t, \forall k = 1, \ldots, h$;

- $\mathcal{R}(s_i^j) = 3, \forall i = 1, \ldots, t, \forall S_j \in \mathrm{SOL}(\mathcal{I})$;
- $\mathcal{R}(s_i^j) = 2, \forall i = 1, \ldots, t, \forall S_j \notin \mathrm{SOL}(\mathcal{I})$.

It is easy to see that $G_\mathcal{R}$ is connected. Moreover the interference experienced by the devices is the following:

- $|I_{\mathcal{R},\mathrm{in}}(u_i)| = \ell, \forall i = 1, \ldots, t$;
- $|I_{\mathcal{R},\mathrm{in}}(s_i^j)| = |S_j| + 2 \leq h + 2, \forall i = 1, \ldots, t$;
- $|I_{\mathcal{R},\mathrm{in}}(v_i^k)| \leq |\mathrm{SOL}(\mathcal{I})|, \forall i = 1, \ldots, t$;
- $|I_{\mathcal{R},\mathrm{in}}(\nu)| = t|\mathrm{SOL}(\mathcal{I})|$;

which implies $I_\mathrm{in}(\mathcal{R}) = |I_{\mathcal{R},\mathrm{in}}(\nu)| = t|\mathrm{SOL}(\mathcal{I})|$.

Now in order to prove the lemma, it remains to prove that any feasible range assignment $\mathcal{R}$ such that $G_\mathcal{R}$ is connected can be used for building a solution for the instance $\mathcal{I}$ of the SCP with size not exceeding $I_\mathrm{in}(\mathcal{R})/t$. First from $\mathcal{R}$, we define a new feasible range assignment $\mathcal{R}'$ such that $I_\mathrm{in}(\mathcal{R}') \leq I_\mathrm{in}(\mathcal{R})$.

We start by assigning $\mathcal{R}'(v) = \mathcal{R}(v)$ to every station v. $\forall i = 1, \ldots, t$, let $U_i \subseteq V_i$ be the set of stations in V_i interfering station ν, i.e., $U_i = \{v \in V_i \mid r(v, \nu) \leq \mathcal{R}(v)\}$. For each $v_i^k \in U_i$, choose a set $S_j \in \mathcal{S}$ containing object o_k and set $\mathcal{R}'(s_j) = \mathcal{R}'(v_i^k) = 3$. Moreover, if $u_i \in U_i$, then let $\mathcal{R}'(u_i) = 2$. Finally, $\forall i, j$, if $\mathcal{R}'(s_i^j) > 3$ then set $\mathcal{R}'(s_i^j) = 3$. Notice that $G_{\mathcal{R}'}$ is connected and the number of stations interfering with ν is not greater than $|I_{\mathcal{R},\mathrm{in}}(\nu)| = \sum_{i=1}^{t} |U_i|$. Since, for any device different from ν, the interference experienced is not greater than $h + 2$, then we have that $I_\mathrm{in}(\mathcal{R}) = |I_{\mathcal{R}',\mathrm{in}}(\nu)| \geq t$.

To conclude the proof, $\forall i = 1, \ldots, t$, let $U_i' \subset V_i$ be the set of stations in V_i interfering station ν in $\mathcal{R}'$, i.e., $U_i' = \{v \in V_i \mid r(v, \nu) \leq \mathcal{R}'(v)\}$. The solution for the instance $\mathcal{I}$ of the SCP is defined as the family of all the sets S_j such that $s_{i*}^j \in U_{i*}'$, where $i^* = \arg\min_{i \in \{1, \ldots, t\}} |U_i'|$. The claim now follows from the inapproximability of the SCP [6] and from the fact that $|U_i'| \leq \frac{I_\mathrm{in}(\mathcal{R})}{t}$. □

By combining the techniques used in the proof of Proposition 4 together with the ones used in the proofs of Propositions 2 and 3, we can prove the following:

Proposition 5. MINMAXRIP *for strong connectivity and for broadcast is not approximable within a factor better than* $c \ln n$, *for some constant* $c > 0$, *unless* $\mathsf{NP} \subseteq \mathsf{DTIME}(n^{\mathcal{O}(\log \log n)})$, *even if* r *is symmetric.* □

Therefore, MINMAXRIP is hard to approximate for any of the considered connectivity predicates. Concerning approximability results, very few is known [18], and thus we conclude by emphasizing the need to concentrate future research efforts along this line.

References

1. M. Burkhart, P. von Rickenbach, R. Wattenhofer, and A. Zollinger, Does topology control reduce interference?, *Proc. of the 5th ACM Int. Symp. on Mobile Ad Hoc Networking and Computing (MOBIHOC'04)*, 9–19, 2004.

2. G. Călinescu, S. Kapoor, A. Olshevsky, and A. Zelikovsky, Network lifetime and power assignment in ad-hoc wireless networks, In *Proc. of 11th European Symp. on Algorithms (ESA'03)*, LNCS 2832, 114–126, 2003.
3. I. Caragiannis, C. Kaklamanis, and P. Kanellopolous, Energy-efficient wireless network design. To appear in *Theory of Computing Systems*.
4. J. Cheriyan and R. Ravi, Lecture Notes on Approximation Algorithms for Network Problems.
5. A. Clementi, P. Crescenzi, P. Penna, G. Rossi, and P. Vocca, On the complexity of computing minimum energy consumption broadcast subgraphs, In *18th Annual Symp. on Theor. Aspects of Comp. Sc. (STACS'01)*, LNCS 2010, 121–131, 2001.
6. U. Feige, A threshold of $\ln n$ for approximating set cover, *Journal of the ACM*, 45(4):634-652, 1998.
7. M. Fussen, R. Wattenhofer, and A. Zollinger, On interference reduction in sensor networks, Technical report, ETH Zürich, Dept. of Computer Science, 2004.
8. L. Jia, R. Rajaraman, and C. Scheideler, On local algorithms for topology control and routing in ad-hoc networks, *Proc. of the 15th Symposium on Parallel Algorithms and Architectures (SPAA'03)*, 220–229.
9. L.M. Kirousis, E. Kranakis, D. Krizanc, and A. Pelc, Power consumption in packet radio networks, *Theoretical Computer Science*, 243:289–305, 2000.
10. L. Li, J.Y. Halpern, P. Bahl, Y.-M. Wang, and R. Wattenhofer, Analysis of a cone-based distributed topology control algorithm for wireless multi-hop networks, *Proc. of the 20th Symp. on Principles of Distributed Computing (PODC'01)*, 264–273.
11. N. Li, J. Hou, and L. Sha, Design and analysis of an MST-based topology control algorithm, *Proc. of the 22th IEEE Conf. on Computer Communications (INFOCOM'03)*, 1702–1712, 2003.
12. X.-Y. Li, G. Calinescu, and P.-J. Wan, Distributed construction of planar spanner and routing for ad-hoc networks, *Proc. of the 20th IEEE Conf. on Computer Communications (INFOCOM'01)*, 2002.
13. W. Liang, Constructing minimum-energy broadcast trees in wireless ad-hoc networks, In *Proc. 3rd ACM Int. Symp. on Mobile ad-hoc Networking and Computing (MOBIHOC'02)*, 112–122, 2002.
14. F. Meyer auf der Heide, C. Schiundelhauer, K. Volbert, and M. Grunewald, Energy, congestion and dilatation in radio networks, *Proc. of the 14th ACM Symp. on Parallel Algorithms and Architectures (SPAA'02)*, 230–237, 2002.
15. T. Moscibroda and R. Wattenhofer, Minimizing interference in ad-hoc and sensor networks, *Proc. of the Joint Workshop on Foundations of Mobile Computing (DIALM-POMC'05)*, 24–33, 2005.
16. R. Ramanathan and R. Hain. Topology control of multihop wireless networks using transmit power adjustment, *Proc. of the 19th IEEE Conf. on Computer Communications (INFOCOM 2000)*, 404–413, 2000.
17. G. Rossi, *The Range Assignment Problem in Static Ad-Hoc Wireless Networks*, Università di Roma "Tor Vergata", Ph.D. Thesis, 2003.
18. P. Von Rickenbach, S. Schmid, R. Wattenhofer, and A. Zollinger, A robust interference model for wireless ad-hoc networks, *5th International Workshop on Algorithms for Wireless, Mobile, Ad-hoc and Sensor Networks (WMAN'05)*, Denver, Colorado, USA, April 2005.

A Context Interpretation Based Wireless Sensor Network for the Emergency Preparedness Class of Applications*

Azzedine Boukerche[1], Regina B. Araujo[1,2], and Fernando H.S. Silva[2]

[1] PARADISE Research Laboratory
SITE, University of Ottawa, Canada
[2] Computer Science Department,
Federal University of São Carlos, Brazil
`boukerch@site.uottawa.ca,`
`{regina, fernando_silva}@dc.ufscar.br`

Abstract. Emergency Preparedness is one of the most appealing classes of applications for context-aware wireless sensor networks (WSN). In such environments, contexts can be captured and interpreted in the WSN application layer to help preventing, fighting, rescuing and checking against fire, explosions, leaking of toxic gases etc. In this paper, we show a Wireless Actor and Sensor Network (WASN) that can be used to interpret simple and complex contexts. We present an efficient technique that make use of actors to aggregate sensor events, eliminate ambiguities and redundancy and realize context interpretations. Each actor of the WASN is configured with rules that are determined by the application in the network configuration phase. Context information is exchanged among actors and sink(s) of the WASN through the publish/subscribe paradigm based on specific topics. We present a set of simulation of experiments to evaluate the performance of our protocols for wireless sensor networks.

1 Introduction

Emergency Preparedness is one of the most appealing classes of applications for context-aware wireless sensor networks (WSN). In such environments, contexts can be captured and interpreted in the WSN application layer to help preventing, fighting, rescuing and checking against fire, explosions, leaking of toxic gases etc. In a fire situation, for instance, it is very important to have a reliable, fine-grain monitoring of the physical environment so that it is possible for the rescue teams to understand what is happening and take the best emergency actions on the environment, preventing lives and patrimony losses. However, in order to understand what is going on in the environment and take the right actions, it is necessary to interpret the collected events.

---

* This work is partially supported by NSERC, Canada Research Chair Program, EAR Research Award, and OIT/Ontario Distinguished Researcher Award. Regina Araujo was a Visiting Scientist at PARADISE Research Laboratory, University of Ottawa, Canada, and CNPq (Brazil).

Accurately [12,13]. Adding new entities named actors to the WSN provides additional advantages to the applications, such as automatic reaction, in real-time, to the events collected by the sensor nodes. The actors have higher cost, with better processing, communication and battery capabilities. In wireless actors and sensors networks (WASN), sensors collect information from the physical world, while the actors process this information in order to make decisions and take actions on the physical environment [2][3]. Actors can make good candidates for the interpretation of more complex contexts in context-aware physical environments for emergency preparedness class of applications. Context is defined here as any information about the physical environment that is important to the application. Sensors capture events. Contexts are composed of events that are related to each other. Contexts can be very simple when they are composed by a single event, such as the actual temperature, actual air pressure, smoke presence etc., or more complex when composed by two or more inter-related events, such as fire occurrence (a combination of the following inter-related events: "high temperature" and "presence of carbon monoxide"). When the relation among the events is temporal, contexts as complex as "why a certain situation happened in the environment?" could be answered through the result of a more complex interpretation that demand the ordering of the events. This paper presents a WSAN that can interpret from simple to complex contexts. In this solution, actors are used as more powerful nodes to aggregate sensor events, eliminate ambiguities and redundancy and realize context interpretations. For that, each actor is configured with rules that are determined by the application (in the network configuration phase). The interpreted contexts can be used by the actors as triggers to actions on the physical environment, such as switching sprinklers on, automatic elevation of environmental temperature etc. Publish/Subscribe mechanism based on topics [8] is used to make interpreted contexts available to other actors or to the sink(s) of the WASN.

The paper is organized as follows: Section 2 discusses context interpretation in WASNs. The importance of time in the relation among events and the causes that lead to delays in a WASN is described in section 3. Related work is described in Section 4. Simulations performed, metrics used and obtained results for context interpretation are discussed in Section 5, followed by Conclusions.

2   Context Interpretation in WANSs

When thinking about context interpretation, one of the first concerns is on which entity of a WASN the interpretation should be carried out. Eligible entities include: the sink; a usually randomly selected sensor node (also called aggregator); and the actor. The sink serves as an interface between the WASN and the application - an input point for the subscription to the network published contexts and also a sensor data aggregator. In [2] two WASN architectures were defined, in which the first one, named Semi-Automatic WSN is the traditional WSN architecture - when a sensor node detects an event, it transmits their reading to the sink. The sink is responsible for interpreting them and, when necessary, triggering the actors. In the second Automatic Architecture, captured events are sent to the actors, which are responsible for interpreting and, if necessary, triggering actions. The Semi-Automatic WSN mirrors a centralized interpretation, in which the sink is the central entity responsible for coordinating all the actors. In the Automatic Architecture, a distributed interpretation occurs in every actor, which is responsible for interpreting the events under its action region. The advantages of a decentralized solution are important to emergency preparedness class of applications and are well known [2]: (a) *low latency*, since the captured events are transmitted to actors that are closer to the event site than the sink; (b) *larger network lifetime,* because data is not routed to the sink, avoiding rapid waste of energy of nodes next to the sink. Although data aggregation can reduce energy consumption in the nodes next to the sink, failure rate in these nodes is still larger than the other nodes in the WSN. Similarly, in the decentralized interpretation it could be thought that there is a larger consumption of energy in the nodes next to the actors. However, the actors are in larger number than the sink(s). Thus, they have more nodes next to them than the sink. Also, there is a big chance that different actors are triggered for each event, so that the nodes are alternated for each event. As a result, decentralized interpretation sustain a larger network lifetime that the centralized one. The use of aggregator nodes is not a good solution when we have actors available. Actors have better processing power and memory capacity allowing more complex interpretations. With better transmission capacity and more energy, actors minimize sensor node energy consumption. Moreover, the use of aggregators demands unnecessary procedures when actors are used, such as: selection of next aggregators; sending of triggering commands to the actors, etc. Thus, actors seem to be an appropriate entity to process context interpretation in WASNs. Context interpretation can be done through artificial intelligence techniques (from models based on rules, case based decision systems, and decision trees to neural networks). Other techniques from different knowledge areas can also be used, such as: graph theory and information and automata theory. The solution presented in this paper for context interpretation in WASNs is as illustrated in figure 1. Context interpretation is decentralized, based on rules (from simple to more complexes, depending on the application). Once contexts are interpreted, they may trigger actions by the actors. Actors can also publish these contexts that are made available to other actors or sink(s) that subscribed to receive them. Examples of interpreted contexts include (from simple to more complex contexts): Peter is in sector A, room 3; Temperature of sector B is 60_oC; There is a fire in sector C; The airplane left wing is broken because of a crack caused by high pressure on that area, etc. How rules are implemented in the actors (XML schemes for rules description) is detailed in [4]. In our solution, the Publish/Subscribe method based on topics is used by actors to publish their contexts and by sinks and other actors to subscribe to the contexts they are interested in. How notifications are sent to subscribers and topics are implemented are described in [4].

2.1 Dealing with Redundancy and Ambiguities

Ambiguities and redundancy are common issues in WASNs that generate several asynchronous events as they occur in the physical environment and are delivered to a corresponding actor. In order to reduce these issues, policies can be created by the application, which are followed by actors when they receive events from the sensors aggregated to them. Examples of policies include: arithmetic average (for instance, in a total of six sensors, four inform region temperature = 60 °C and the other two inform temperature = 48 °C. If arithmetic average is requested by the application, the notification to actors and sink(s) that subscribed to that information will be average temperature = 56 °C); Moda (value to be considered is the value that occur more frequently), average (for instance, discard the most discrepant values – the lowest and highest); and etc.

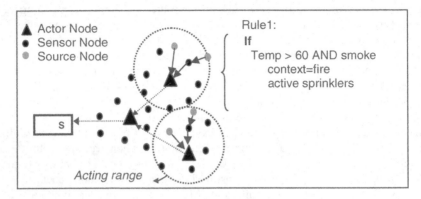

Fig. 1. Context Interpretation in Actor Nodes

3   Interpretation and Ordering of Correlated Events

Events that are captured by sensors may (or not) be time related to each other (correlation). When two or more events are not correlated, each incoming event can be treated independently. However, when events are correlated, a mechanism for capturing and interpreting these events, jointly, is necessary. Event correlation is becoming an important service for reactive distributed applications, such as the WASNs. It combines the information collected from individual devices into a higher-level knowledge, adding value to data management, scalability and performance. Events can be related to other events not only in terms of context but also in terms of the time each of them occurred. This means that the order in which events were "sensed", when they were sensed, and the elapsed time between each event can be important for the detection of emergency situations. Event ordering can be a solution to the temporal correlation [10].

3.1   Event Ordering

To illustrate the importance of event ordering, consider the following example: A physical environment is covered with a wireless sensor network, with different types of sensors (temperature, smoke detection, fire detection, object/human location etc). The physical environment is monitored through a 3D interface that mimics the physical environment, showing, in a fine-grain monitoring, what is happening in the physical environment – this 3D interface can be visualized through a lap-top or a PDA (Personal Digital Assistant) [6]. In a fire situation, a person inside a physical environment in fire tries to find an exit. The rescue team manager fire fighter tries to locate the person from outside the physical environment, by monitoring her location through the 3D interface, in order to plan the rescue. As the person searches for the exit, sensors capture her location and send it to the manager fire fighter. Figure 2 (a) shows the correct trajectory of this person, who moves in the sensors sequence *1, 2* and *3*. However, a delay in sending the monitoring message for sensor *2*, makes the sensor *3* message to arrive before sensor *2* message, resulting in an incorrect interpretation of that person's trajectory, as shown in Figure 2 (b). This could lead the rescue team to go to the opposite direction to the real movement of this person. Therefore, when monitoring messages are received, an interpretation can not be triggered right away, since messages that arrived before those that just arrived, can have suffered delays and still be in transit in the network. The ordering of messages is necessary because they may suffer delay along their trajectory from the source node to the sink node. Given the communication nature of WSNs, it is not possible to assume that the order of the collected events from different parts of the system reflects the order in which events occurred or were sent without taking certain precautions. Thus, in order to present a consistent and coherent view of the monitored events an ordering of the monitoring messages is necessary.

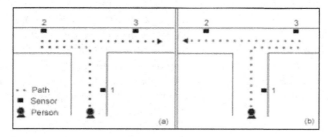

Fig. 2. Ordering of events: (a) Correct ordering; (b) Incorrect ordering due to message 2 delay

3.2 Ordering by Confirmation (OBC) – A Low Latency and Energy Aware Ordering Algorithm for WSANs

In general, only events that are generated by a subgroup of network sensors should be delivered in a temporal order. WSAN networks usually have already such subgroups associated to them. Thus, the ordering of events is restricted to these subgroups, in which all ordered subgroups set up a whole network ordered. Algorithms developed for WSNs do not take advantage of the additional capability provided by more powerful nodes, such as the actors. We proposed the OBC algorithm in which actors are major entities that order the events in an efficient, fast and energy aware way. The OBC algorithm, similarly to the previous algorithms described, also requires a pre-configuration of the sensor network. This pre-configuration involves: physical clock synchronization; discovery of routes and paths; and assignment of FIFO channels among all nodes of the network. The OBC algorithm is fully described in [7].

4 Related Work

A context detection mechanism called PROCON is described in [1]. In PROCON, context decisions are made in a distributed way, through the cooperation of nodes called "event headers", which are connected via a context overlay network. The "event headers" are similar to other sensor nodes of the network – they have the same processing power and energy and transmission capabilities. Because of the typical low transmission power of the sensor nodes, the communication among "event headers", is multi-hop, using the other sensors of the network. It is well known that multi-hop communication can lead to larger delays since messages are exchanged among more sensors, consuming also more energy. Differently from PROCON, our solution uses more powerful actor nodes as aggregator and context interpreter nodes. When other actors or sink(s) need a context from a particular actor, they can subscribe to that context. Thus, only the nodes that subscribed to the context will receive it. Solutions, such as the ones presented in [9], [10] and [5], use Query-Based protocols for getting desired information from a WSN. These solutions basically view a WSN as a distributed data base, in which relational data base relational queries like are sent to all sensor nodes. The disadvantages of such centralized solutions are well known: data travels along longer multi-hop paths spending more energy, reducing the overall network lifetime. Moreover, centralized solutions typically present larger latency. A WSN architecture is described in [11], SINA, which provides monitoring by assigning tasks to sensor nodes that can be queried for their status. SINA adopts a data centric naming scheme and has three functional components: hierarchical clustering, attribute-based naming and location awareness. A Sensor Query and Tasking Language (SQTL) was created that can interpret simple declarative queries and support query-based, periodic and event-driven types of applications. COUGAR [5] and TinyDB [9] enable users´ input queries at the server though a simple, SQLlike language that describes the data they wish to collect and how they wish to combine, transform, and summarize it. They use the in-network concept, in which data is aggregated as it moves towards the sink. Results from queries can be stored to be used in future queries. There are not many reported works in the literature on WASN context interpretation. As these networks become more powerful, more complex interpretations will be able to be processed and more autonomy will be given to the networks to act upon the physical environments.

5   Simulation Experiments

This section presents the performance analysis for the context interpretation proposed in section 2. For the context interpretation, simulations were realized in order to show how the decentralized solution proposed performs better than a centralized one. Since WASNs are being considered in this work, the network is divided in subgroups of nodes and the simulation is performed on a WASN subgroup. The simulation scenario consists of a sensor field containing subgroups of sensors with different number of nodes – at every sub-group there is an actor, which is responsible for coordinating the ordering, actuation and context interpretation of that area. These nodes are randomly placed in the field. Each sensor node is deployed at a distance from the actor. A percentage of the sensor nodes is randomly chosen to produce events. The data rate for each source node is also determined randomly. Each value measured was taken from a mean of 20 simulations. It is assumed, for the simulation of both algorithms, that the paths set up by the routing algorithm are perfect (paths do not break). A confidence of 95% was achieved with a 4% error level. Table 1 lists the simulation parameters used.

Table 1. Simulation Parameters

Parameters	Values
Simulation area	160x160m
Simulation time (s)	24
Number of nodes	100-240
Number of Source Nodes	20
Source data rate (events/s)	Random
Radio range (m)	20
Transmit Energy (mW)	14.88
Receive Energy (mW)	12.50
Dissipation in Idle (mW)	12.36
Dissipation in Sleep (mW)	0.016
One Hop Delay(ms)	20
Number of Actors	4

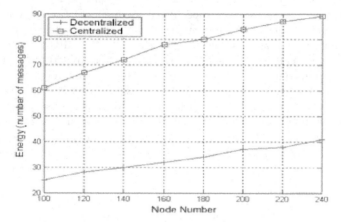

Fig. 3. Energy dissipation vs sensor nodes

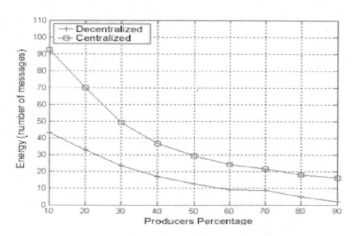

Fig. 4. Energy dissipation vs event rate

The algorithm was analyzed based on two metrics: *event delivery delay* - the amount of time elapsed between the detection of an event and its notification to the application; and *dissipated energy* - energy dissipation is important because the nodes are devices with limited power source and also because energy dissipation can have a large impact on delay. Figure 3 shows the energy dissipation per message for sensor networks with different densities. It can be seen that the increase in the number of sensor nodes will not influence directly the average energy dissipation for temporal confirmation and posterior interpretation because it will not be created **longer** paths but **a larger number of paths**. However, by increasing the number of sensors, the number of source nodes that generates messages will also be increased since the number of source nodes is a percentage of the sensor nodes number. Figure 4 shows the average energy dissipation of the algorithm when different rates of events are generated. For our decentralized solution that embeds the OBC event ordering algorithm, which confirms more messages all at once as part of the ordering process, latency is decreased, therefore reducing the dissipated energy for each sent message. Figures 5 and 6 show the latency by varying, respectively, the number of sensor nodes and the event rate. For our decentralized solution, the latency decreases - again mainly due to the fact that more messages are confirmed all at once by the OBC algorithm. For the OBC, latency is estimated by the number of messages transmitted for the event confirmation, through the largest path and also by sending the event to the actor. Thus, by increasing the nodes, the size of the path does not increase (this only happens when the dimension of the sensors field is increased) – what increases is the amount of paths. For our solution of decentralized context interpretation in WASNs, the graphics showed the better results obtained. In summary, this is due to the fact that when interpretation is performed in a decentralized way, the number and size of paths for the transmission and confirmation messages are reduced implying in reduced energy consumption and reduced latency.

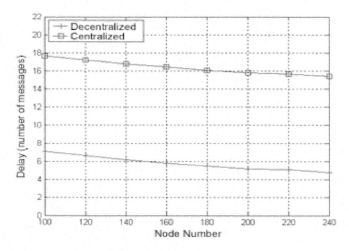

Fig. 5. Network delay vs sensor nodes

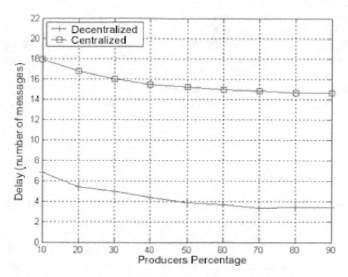

Fig. 6. Network latency vs event rate

6 Conclusions

Wireless Actor and Sensor Networks (WASNs) are increasingly being deployed for fine-grain monitoring of physical environments subjected to critical conditions such as fire, leaking of toxic gases and explosions. This paper presented a solution to context interpretation through a WASN, in which actors are used as more powerful nodes to aggregate sensor events, eliminate ambiguities and redundancy and realize context interpretations. Each actor of the WASN is configured with rules that are determined by the application in the network configuration phase. Our solution was compared to a centralized solution, in which events are collected by the sink. Results show that when interpretation is performed in a decentralized way, the number and size of paths for the transmission and confirmation messages are reduced leading to reduced energy consumption and latency. Because the order in which events are received can determine the correct interpretation of what is going on in the physical environment being monitored, a novel protocol for event ordering in WASNs, OBC (Ordering by Confirmation), was also introduced (details of simulation results were reported in [7]). The combined solution of having actors to process context interpretation and event ordering can be a good alternative for emergency preparedness applications where lives and patrimony are at risk [12].

References

1. Ahn, S. and Kim, D. "System Proactive Context-Aware Sensor Networks". In European Workshop on Wireless Sensor Networks (EWSN) , Zurich, Switzerland, 13-15, 2006.
2. Akyildiz, I.F. and Kasimoglu, I.H. "Wireless sensor and actor networks: research challenges", Ad Hoc Networks, Elsevier, Volume 2, Issue 4, October 2004, p. 351-367.

3. Akyildiz, I.F., Su, W., Sankarasubramaniam, Y., Cayirci, E. "Wireless sensor networks: A survey", Computer Networks, March 2002, p. 393-422.
4. Araujo, R.B., Kudo, N. T. and Michelotti, G. "Context interpretation through the Publish/Subscribe Mechanism based on Topics". Technical report, CS/UFSCar, April, 2005.
5. Bonnet, P., et al. Towards Sensor Database Systems, In Proceedings of the Second International Conference on Mobile Data Management, 2001.
6. Boukerche, A., Lopes, Araujo, R.B. "Recording and Playing 3D Collaborative Virtual Environment Simulations as Continuous Media for Critical Security Applications". In 4th I2TS´2005, Florianopolis, Brazil, October, 2005.
7. Boukerche, A., Silva, F. H. S. and Araujo, R.B. – "An Event Ordering Algorithm that Extends the Lifetime of Wireless Actor and Sensor Networks". In The 8th ACM/IEEE Int'l Symposium Modeling, Analysis and Simulation of Wireless and Mobile Systems (MSWiM'05), Montreal, Canada.
8. Eugster, P.T., Felber, P.A., Guerraoui, R., Kermarrec, A.M. "The Many Faces of Publish/Subscribe". ACM Computing Surveys, p.114-131, Junho/2003.
9. Madden, S. et al. "TinyDB: An acquisitional query processing system for sensor networks". In Transactions on Database Systems (TODS), 2005Hayton, R. "OASIS: An Open Architecture for Secure Interworking Services". Phd thesis, CambridgeU, 1996.
10. Mühl, G., Ulbrich, A. and Ritter, H. "Content Evolution Driven Data Propagation in Wireless Sensor Networks". Em proceedings MuUlRi:2004:Funnel, p. 55-57, 2004.
11. Shen, C., Srisathapornphat, C. and Jaikaeo,C., "Sensor Information Networking Architecture and Applications". IEEE Personal Communications, Aug. 2001, pp. 52-59.
12. A. Boukerche, "Handbook of Algorithms for Wireless Networking and Mobile Computing", CRC/Hall Chapman , 2005
13. A. Boukerche, R. W. N. Pazzi and R. Araujo, " Fault-tolerant wireless sensor network routing protocols for the supervision of context-aware physical environments", Journal of Parallel and Distributed Computing, V. 66, Issue 4, April 2006, Pages 586-599.

Adaptive Initialization Algorithm for Ad Hoc Radio Networks with Carrier Sensing*

Jacek Cichoń, Mirosław Kutyłowski, and Marcin Zawada

Institute of Mathematics and Computer Science
Wrocław University of Technology
Poland

Abstract. We propose an algorithm for coordinating access to a shared broadcast channel in an ad hoc network of unknown size n. We reduce the runtime necessary to self-organize access to the channel over the previous algorithm of Cai, Lu and Wang. The runtime of this algorithm is $O(n)$, our work is to improve the constant factors involved. Apart from experimental evidence of algorithm quality, we provide a rigorous probabilistic analysis of its behavior.

Keywords: Ad hoc network, radio channel, initialization, carrier sensing.

1 Introduction

We consider an ad hoc network consisting of processing units, called *stations* and communicating through a shared radio channel. We assume that some of these stations, that we call *active*, wish to transmit data. However, it is unpredictable which stations are active. We also assume that the time is divided into slots and clocks of all stations are synchronized.

The problem is to design a protocol so that each of active stations gets its time slot for transmission, while the total time used should be kept as small as possible.

The scenario that we proceed is that active stations receive the subsequent numbers $1, \ldots, m$ (initialization procedure), and afterwards transmit data: station i during the ith time slot. For the initialization procedure neither infrastructure nor central control unit is established, that means the active stations have to self-organize themselves.

We consider single-hop networks: a message sent by one station can be received by all stations. Each station is capable to sense the *carrier*. It means that a station can identify the channel status as either *busy* (some station is sending signals) or *idle* (no station is sending a signal). While carrier sensing is a standard feature of modern wireless devices [10], the signal is received with a certain delay due to the distance between the stations. So if a station finds that the channel is idle, it is not necessarily true, because in the meantime some station might have started sending a signal. This makes it hard to design a protocol

---

* Partially supported by the EU within the 6th Framework Programme under contract 001907 (DELIS).

that coordinates who will transmit within a slot. On the other hand, if two or more stations broadcast messages within the same time, then a *collision* occurs i.e. interference of signals makes the messages unreadable.

For the sake of simplicity we assume that there is a *leader* in the network that will send acknowledgments (*AP* for short) at the end of each slot. We skip the details of leader election (for details see for instance [2]).

1.1 Previous Algorithms

Nakano and Olariu [8] proposed the following basic algorithm that works if the number of stations n is known. Within a time slot each active station sends a message with probability $1/n$. If exactly one station has transmitted, then it receives the smallest unused ID. This procedure is repeated by the stations having no ID until no such active station is left. The protocol terminates in time $e \cdot n + O(\sqrt{n \log n})$ with probability exceeding $1 - \frac{1}{n}$.

The second protocol from [8] works when the number of stations is unknown. The stations assign themselves into different groups, the groups are split and finally each station is in a different group. At the beginning all stations are in the same group. The protocol terminates in time $\frac{10n}{3} + O(\sqrt{n \log n})$ with probability exceeding $1 - \frac{1}{n}$. In [9], a hybrid method for known n is proposed. It has runtime $\approx 2.15 \cdot n$. A serious drawback of protocols [8,9] is the assumption that stations which are transmitting can determine if a collision has occurred. However, in the practical technical setting it is hard to achieve.

Cai, Lu and Wang [2] proposed an initialization algorithm that utilizes carrier sensing. Each time slot is divided into the following minislots: initial period (IP), transmission period (TP), acknowledgment period (AP). In period IP one station is chosen (with high probability), then period TP is used for sending the workload message of this station, and finally during period AP correctness of transmission is acknowledged to the station that has transmitted.

Each active station chooses a backoff period uniformly at random within minislot IP. In the moment indicated by the backoff, the station checks, whether the communication channel is busy. If carrier is detected, then the station is not initialized and does not transmit within this slot. Otherwise, it starts to send carrier signal until the end of TP. During TP it sends a workload message. Of course, it may happen that after the station with the smallest backoff starts sending carrier signal, the second station monitors the channel and detects no carrier, due to signal propagation delay. In this case both stations will transmit and this slot will be useless.

If period IP is long, then probability of collision is small (exact values depend on actual propagation delays), but the time overhead of period IP is large. If we reduce the length of IP period, then the overhead is reduced, but the number of trials increases due to more frequent collisions. The runtime achieved depends on propagation delay (see Table 2).

Time Settings of the Scheme. We use notation from paper [2]: duration of a slot is denoted by α (it must be set according to technical reality, see [4,10]). At the end of each slot there is a fixed period of length 1 devoted to acknowledgments, the initial period of length $\alpha - 1$ can be utilized by the algorithm. Let λ be the maximum propagation delay of a signal in the network [2] normalized to the length of acknowledgements. The transmission radius for the stations in the single-hop ad hoc network is limited in practical situations, it is usually less than 1000 meters, so $\lambda \ll 1$. Let $\delta = \frac{\lambda}{\alpha - 1}$ be the relative length of propagation delay.

1.2   Our Contribution

We design a new initialization algorithm for an unknown number of active stations. If their number is n, then the expected runtime achieved is $(1 + C) \cdot n$ (see Section 3.5), where C is a parameter depending on propagation delay in a quite complicated way described in Section 3.5. Our scheme outperforms [2], concerning the runtime constant in front of n.

2   Algorithm Description

Now we give an informal description of our algorithm; its pseudocode is given in Fig. 1. The algorithm executes a number of stages, during each stage a number of stations receive their consecutive numbers. Each stage consists of two parts, first a group of stations called *final winners* is chosen, then the next consecutive ID's are assigned to them.

For the first part a number of rounds is executed, during each round the stations that are still *winners* toss a biased coin. Initially all participating stations are winners. Those stations who throw tails become losers, while those who throw heads remain as winners (and send carrier signal during the corresponding slot). These steps are repeated until all stations have lost (no carrier is acknowledged by the leader).

The stations which have been winners in the last round, that we call *final winners*, execute a procedure based on the algorithm from [2]: each of, say m, final winner generates a backoff value $\gamma_i \in [0, 1]$ uniformly at random and senses the channel at time γ_i within the slot. The second version of the algorithm uses here a non-uniform distribution. If the channel is busy, then the station is not trying to get ID in this slot. Otherwise it starts sending carrier signal. Therefore a single station transmits in the slot only if $\gamma_{(2)} - \gamma_{(1)} > \delta$, where $\gamma_{(1)}, \ldots, \gamma_{(m)}$ denote the backoff values after sorting them. The probability of this event is $(1 - \delta)^m$ (see [3]). This is repeated until all current winners have transmitted successfully (this can be detected by the fact that there is no carrier signal sent in a slot). Then we start a new stage with the same stations except for the final winners.

3   Analysis of the Algorithm

We assume that the reader is familiar with the concepts of complex functions (the reader can also consult the literature, see for instance [1]). By i we denote the imaginary unit and by $\Re(z)$ we denote the real part of the complex number z.

Initialization-Algorithm

```
var     N: integer                      init 1
        id : integer ∪ {undef}          init undef
        p* : const float
```

```
1.      forever
2.          repeat
3.              wait for begin of a slot
4.              γ := rand(0, 1)
5.              if γ ≤ p* then send ⟨msg⟩
6.          until γ > p*
7.          wait for begin of a slot
8.          receive ⟨msg⟩, AP message from the leader
9.          if msg = SILENCE then
10.             repeat
11.                 wait for begin of a slot
12.                 if version = 1 then γ := rand(0, 1)
13.                 if version = 2 then γ := √rand(0, 1)
14.                 channel := sense the channel at (α − 1)γ
15.                 if channel = idle then transmit ⟨carrier⟩
16.                 receive ⟨msg⟩, AP message from the leader
17.                 if msg = SINGLE then
18.                     if channel = idle then // my packet has been transmitted
19.                         set id := N and leave
20.                     N := N + 1
21.                 end
22.             until msg = SILENCE
23.         else
24.             repeat
25.                 wait for begin of the slot
26.                 receive ⟨msg⟩, AP message from the leader
27.             until msg = SILENCE
28.             repeat
29.                 wait for begin of the slot
30.                 receive ⟨msg⟩, AP message from the leader
31.                 if msg = SINGLE then N := N + 1
32.             until msg = SILENCE
33.         end
34.     end
```

Fig. 1. Pseudo-code of our algorithm executed by a single station, parameter *version* indicates the first or the second version of choosing backoff values

By $\mathrm{B}(x, y)$ we denote the Euler beta function defined by $B(x, y) = \frac{\Gamma(x)\Gamma(y)}{\Gamma(x+y)}$. We will use the fact that the function $B(n + 1, z)$ is analytic everywhere except for $z = 0, -1, -2, \ldots$, its residue at $z = -k$ equals $\mathrm{Res}_{z=-k} \mathrm{B}(n+1, z) = \binom{n}{k}(-1)^k$.

By $\mathrm{H}_n = \sum_{k=1}^{n} \frac{1}{k}$ we denote the n-th harmonic number. Let us recall that $\mathrm{H}_n \simeq \log(n) + \gamma + \frac{1}{2n}$ where $\gamma \approx 0.5772$ is the Euler-Mascheroni constant.

3.1   The Number of Final Winners

First we calculate the probability distribution of the number of winners after k tosses of a biased coin. Let p be the probability of throwing a head and $X(n)$ denotes the number of winners after each of n stations tosses a coin only once. Obviously, $X(n)$ follows the binomial distribution $\mathbf{B}(n,p)$. Let $\mathcal{B}_n(z) = \mathbf{E}z^{X(n)} = \sum_k P\left[X(n) = k\right] \cdot z^k$. It is straightforward to check that $\mathcal{B}_n(z) = (1-p+pz)^n$. From this we easily deduce the following result.

Lemma 1. *Let $X^k(n)$ be a random variable denoting the number of stations which are winners after k tosses of a biased coin with probability p of throwing heads and let $\mathcal{B}_n^k(z) = \mathbf{E}[z^{X^k(n)}]$. Then $\mathcal{B}_n^k(z) = (1 - p^k + p^k z)^n$.*

Next we calculate the number of final winners.

Lemma 2. *Let $Y(n)$ be a random variable denoting the number of final winners, when we start with n stations. Let $\mathcal{Y}_n(z) = \mathbf{E}z^{Y(n)}$. Then $\mathbf{E}[Y(1)] = 1$, $\mathbf{E}[Y(2)] = \frac{2}{1+p}$ and if $n > 2$, then*

$$\mathbf{E}[Y(n)] = \frac{(1-p)}{p\log(1/p)} + 2\frac{n(1-p)}{p\log(1/p)} \sum_{k=1}^{\infty} \Re\left[\mathbf{B}\left(n, 1 + \frac{2k\pi\mathrm{i}}{\log(p)}\right)\right] \ . \tag{1}$$

Proof. We split the function $\mathcal{Y}_n(z) = \sum_{t=1}^n \Pr[Y(n) = t] \cdot z^t$ into a sum of a sequence of functions $\mathcal{Y}_n^k(z)$, namely we put

$$\mathcal{Y}_n(z) = \sum_{k=1}^{\infty} \mathcal{Y}_n^k(z) \ , \tag{2}$$

where

$$\mathcal{Y}_n^k(z) = \sum_{t=1}^{n} \Pr[X^1(n) > 0, \dots, X^{k-2}(n) > 0, X^{k-1}(n) = t, X^k(n) = 0] \cdot z^t \ .$$

It is easy to check that

$$\mathcal{Y}_n^k(z) = (1 - p^{k-1} + p^{k-1}(1-p)z)^n - (1 - p^{k-1})^n \ . \tag{3}$$

We calculate $\mathbf{E}[Y(n)]$ by computing the first derivative of the function $\mathcal{Y}_n(z)$ at the point $z = 1$:

$$\mathbf{E}[Y(n)] = \mathcal{Y}_n'(1) = \sum_{k=1}^{\infty}(\mathcal{Y}_n^k(1))' = \sum_{k=1}^{\infty} np^{k-1}(1-p)(1-p^k)^{n-1} =$$

$$n(1-p)\sum_{k=1}^{\infty} p^{k-1}(1-p^k)^{n-1} = \frac{n(1-p)}{p}\sum_{k=1}^{\infty} p^k(1-p^k)^{n-1} =$$

$$\frac{n(1-p)}{p}\sum_{l=0}^{n-1}\binom{n-1}{l}(-1)^l\frac{1}{1-p^{l+1}} \ .$$

Let $f(z) = \frac{1}{1-p^{-z+1}}$ be a complex function for $z \in \mathbb{C}$. It is analytic except for $\mathfrak{z}_k = 1 + \frac{2k\pi\mathrm{i}}{\log(p)}$ for $k \in \mathbb{Z}$. If $n > 2$, then we can use the method of treatment of alternating sums of the form $\sum_k \binom{n}{k}(-1)^k a_k$ attributed by D. Knuth (see [7]) to S. O. Rice and we get

$$\sum_{k=0}^{n} \binom{n-1}{k}(-1)^k \frac{1}{1-p^{k+1}} = -\sum_{k=-\infty}^{\infty} \operatorname*{Res}_{z=3k} \mathrm{B}(n,z)f(z) \ .$$

It is easy to check, that

$$\operatorname*{Res}_{z=3k} \mathrm{B}(n,z)f(z) = \begin{cases} \frac{1}{n\log(p)} & \text{for } k = 0 \ , \\ \frac{\Gamma(n)\Gamma(3k)}{\Gamma(n+3k)\log(p)} & \text{for } k \neq 0 \ , \end{cases}$$

for $k \in \mathbb{Z}$. Since $\overline{\Gamma(z)} = \Gamma(\overline{z})$ and $z+\overline{z} = 2\Re[z]$, we obtain the result claimed. □

3.2 Runtime of the First Part of a Stage

Now we concern the number of rounds after which there are no winners. Formally, the event that after k rounds there is no winner is

$$X^1(n) > 0, \ldots, X^{k-2}(n) > 0, X^{k-1}(n) > 0, X^k(n) = 0 \ . \tag{4}$$

Lemma 3. *Let $T(n)$ be a random variable denoting the number of rounds such that the number of winners becomes 0, when we start with n stations. Then*

$$\Pr[T(n) = k] = (1 - p^k)^n - (1 - p^{k-1})^n \ , \tag{5}$$

$\mathbf{E}[T(1)] = \frac{1}{1-p}$, $\mathbf{E}[T(2)] = \frac{1+2p}{1-p^2}$ *and*

$$\mathbf{E}[T(n)] = \frac{1}{2} + \frac{H_n}{\log(1/p)} + \frac{2}{\log(1/p)} \sum_{k=1}^{\infty} \Re\left[\mathrm{B}\left(n+1, \frac{2k\pi i}{\log(p)}\right)\right] \tag{6}$$

for $n > 2$.

Proof. We use in the proof function $\mathcal{Y}_n^k(z)$ defined in the proof of Lemma 2 and Equation (3):

$$\Pr[T(n) = k] = \Pr[X^1(n) > 0, \ldots, X^{k-1}(n) > 0, X^k(n) = 0] =$$
$$\Pr[X^{k-1}(n) > 0, X^k(n) = 0] = \mathcal{Y}_n^k(1) = (1 - p^k)^n - (1 - p^{k-1})^n \ .$$

Since we have no closed form for $\mathbf{E}[z^{T(n)}]$, we calculate $\mathbf{E}[T(n)]$ step by step:

$$\mathbf{E}[T(n)] = \sum_{k=1}^{\infty} k\left((1 - p^k)^n - (1 - p^{k-1})^n\right) =$$

$$\sum_{k=1}^{\infty} k \sum_{i=0}^{n} \binom{n}{i}(-1)^i((p^k)^i - (p^{k-1})^i) =$$

$$\sum_{i=1}^{n} \binom{n}{i}(-1)^i \left(\sum_{k=1}^{\infty} k(p^i)^k - \sum_{k=1}^{\infty} k(p^i)^{k-1}\right) =$$

$$\sum_{i=1}^{n} \binom{n}{i}(-1)^i \left(\frac{p^i}{(1-p^i)^2} - \frac{1}{(1-p^i)^2}\right) = -\sum_{i=1}^{n} \binom{n}{i}(-1)^i \frac{1}{1-p^i} \ .$$

Let $f(z) = \frac{1}{1-p^{-z}}$ be a complex function for $z \in \mathbb{C}$. It is analytic except for $z_k = \frac{2k\pi i}{\log(p)}$, $k \in \mathbb{Z}$. If $n > 2$, then using Rice method we get

$$\sum_{k=1}^{n} \binom{n}{k}(-1)^k \frac{1}{1-p^k} = - \sum_{k=-\infty}^{\infty} \operatorname*{Res}_{z=z_k} B(n+1, z) f(z) \ .$$

It is easy to check that

$$\operatorname*{Res}_{z=z_k} B(n+1, z) f(z) = \begin{cases} \frac{1}{2} - \frac{H_n}{\log(p)} & \text{for } k = 0 \ , \\ B(n+1, z_k)\frac{1}{\log(p)} & \text{for } k \neq 0 \ , \end{cases}$$

from which we get the required formula. $\square$

3.3   Wasted Slots in the Second Part of a Stage

We analyze now the second part of a stage. Let $L_1(m)$ be a random variable denoting the number of slots in which more than one station transmits i.e. the number of slots that have been wasted for assigning ID's. By definition, $L_1(m)$ is estimated from above by the negative binomial distribution $\mathbf{NB}(m, (1-\delta)^m)$.

Lemma 4. *If $n > 2$, then the expected number of wasted slots $\mathbf{E}[L_1(Y(n))]$ is approximated by*

$$\frac{(1-p)}{(p-\delta)\log(\frac{1}{p})} + \frac{2n(1-p)}{(p-\delta)\log(\frac{1}{p})} \sum_{k=1}^{\infty} \Re\left[\left(\frac{1-\delta}{p-\delta}\right)^{\frac{2\pi i}{\log(p)}} B\left(n, 1 + \frac{2k\pi i}{\log(p)}\right)\right] -$$
$$\left(\frac{1-p}{p\log(\frac{1}{p})} + \frac{2n(1-p)}{p\log(\frac{1}{p})} \sum_{k=1}^{\infty} \Re\left[B(n, 1 + \frac{2k\pi i}{\log(p)})\right]\right) \ . \tag{7}$$

Proof. Let us recall that if $L_1(m) \sim \mathbf{NB}(m, (1-\delta)^m)$, then we have $\mathbf{E}[L_1(m)] = m\frac{1-(1-\delta)^m}{(1-\delta)^m}$. However, m itself is a random variable. We proceed as follows:

$$\mathbf{E}[L_1(Y(n))] = \sum_{k=0}^{\infty} k \cdot \Pr[L_1(Y(n)) = k] =$$

$$\sum_{k=0}^{\infty}\sum_{i=0}^{n} k \cdot \Pr[L_1(i) = k] \cdot \Pr[Y(n) = i] = \sum_{i=0}^{n} \mathbf{E}[L_1(i)] \cdot \Pr[Y(n) = i] =$$

$$\sum_{i=0}^{n} i(\frac{1}{(1-\delta)^i} - 1) \cdot \Pr[Y(n) = i] = \sum_{i=0}^{n} i(1-\delta)^{-i} \Pr[Y(n) = i] - \mathbf{E}[Y(n)] \ .$$

To calculate the sum $\sum_{i=0}^{n} i(1-\delta)^{-i} \Pr[Y(n) = i]$ we define an auxiliary function

$$\mathcal{L}(z) = \sum_{i=0}^{n} i((1-\delta)z)^{-i} \cdot \Pr[Y(n) = i] \ .$$

(Notice that $\sum_{i=0}^{n} i(1-\delta)^{-i} \Pr[Y(n) = i] = \mathcal{L}(1)$.) Then

$$\mathcal{L}(z) = -z\frac{\partial}{\partial z}\sum_{i=0}^{n}((1-\delta)z)^{-i} \cdot \Pr[Y(n) = i] = -z\frac{\partial}{\partial z}\mathcal{Y}\left(\frac{1}{(1-\delta)z}\right),$$

from which, after an easy calculation, we get

$$\mathcal{L}(z) = z\frac{n(1-p)}{(1-\delta)z^2}\sum_{k=1}^{\infty}p^{k-1}\left(1-p^{k-1}+\frac{p^{k-1}(1-p)}{(1-\delta)z}\right)^{n-1}.$$

So we have

$$\mathcal{L}(1) = \frac{n(1-p)}{1-\delta}\sum_{l=0}^{\infty}\binom{n-1}{l}(-1)^{l}\left(\frac{p-\delta}{1-\delta}\right)^{l}\frac{1}{1-p^{l+1}}.$$

Let us consider the complex function $f(z) = \left(\frac{p-\delta}{1-\delta}\right)^{-z}\frac{1}{1-p^{-z+1}}$. It is analytic except for $\mathfrak{z}_k = 1 + \frac{2k\pi i}{\log(p)}$ for $k \in \mathbb{Z}$. If $n > 2$, then using Rice method we get

$$\sum_{k=0}^{n}\binom{n-1}{k}(-1)^{k}\left(\frac{p-\delta}{1-\delta}\right)^{k}\frac{1}{1-p^{k+1}} = -\sum_{k=-\infty}^{\infty}\operatorname*{Res}_{z=\mathfrak{z}_k} B(n,z)f(z).$$

A simple calculation shows that

$$\operatorname*{Res}_{z=\mathfrak{z}_k} B(n,z)f(z) = \begin{cases} \dfrac{1-\delta}{n(p-\delta)\log(p)} & \text{for } k = 0, \\[3mm] B(n,\mathfrak{z}_k)\left(\dfrac{p-\delta}{1-\delta}\right)^{-\mathfrak{z}_k}\dfrac{1}{\log(p)} & \text{for } k \neq 0. \end{cases}$$

Summing up all residues, using Lemma 2 and the equality $\mathbf{E}[L_1(Y(n))] = \mathcal{L}(1) - \mathbf{E}[Y(n)]$ we get the required result. $\qquad\square$

3.4   Second Version

We formulate without proofs some properties of the second version of the algorithm:

Lemma 5. *Let the final winner generate a backoff value $\gamma_i \in [0,1]$ from cumulative distribution function $F(x) = x^2$ for $0 \leq x \leq 1$ and 0 otherwise. Then*

$$\Pr(\gamma_{(2)} - \gamma_{(1)} \geq \delta) = 2n\int_{0}^{1-\delta} x(1-(x+\delta)^2)^{n-1}dx$$

which is approximately

$$P(\delta, n) = (1-\delta^2)^{n-1} - \delta\sqrt{\pi(n-1)} + 2\delta^2(n-1). \tag{8}$$

We consider now the second part of the stage for the second version of the algorithm. Let $L_2(m)$ be the random variable denoting the number of slots in which more than one station transmits - just like $L_1(m)$ for the first version. $L_2(m)$ is approximated by negative binomial distribution $\mathbf{NB}(m, P(\delta, m))$. The following lemma may be proved in a similar way as Lemma 4.

Lemma 6

$$\mathbf{E}[L_2(Y(n))] \leq \delta \sqrt{\pi(n-1)}\mathbf{E}[Y(n-1)] + \delta^2(\pi-1)\,\mathbf{E}[Y^2(n-1)] \ . \qquad (9)$$

3.5   Total Runtime

We begin with a recurrence relation for the expected number of wasted slots in the whole algorithm. Namely we have

Lemma 7. *Let $S(n)$ be the number of overhead slots necessary to assign IDs to all n stations. Then*

$$\mathbf{E}[S(n)] = \mathbf{E}[T(n)] + \mathbf{E}[L_i(Y(n))] + \sum_{r-1}^{n} \Pr[Y(n) = r]\mathbf{E}[S(n-r)] \ . \qquad (10)$$

We omit an easy proof of this lemma. Notice that the expected number of all slots used by our algorithm equals $\mathbf{E}[S(n)] + n$. Let

$$C_a(p, \delta, U) = \max_{a \leq m \leq U} \left\{ \frac{1}{\mathbf{E}[Y(m)]} \right\} \cdot \max_{a \leq m \leq U} \{\mathbf{E}[T(m)] + \mathbf{E}[L_i(Y(m))]\} \ .$$

and

$$D_a(p, \delta, U) = \max_{1 \leq m < a} \mathbf{E}[S(n)]$$

From Lemma 7 we get the following upper estimation:

Theorem 1. $\mathbf{E}[S(n)] \leq C_a(p, \delta, U) \cdot n + D_a(p, \delta, U)$ *for n such that $a \leq n \leq U$.*

Therefore the number $(1 + C_a(p, \delta, U)) \cdot n + D_a(p, \delta, U)$ is an upper bound of the expected number of all slots used by our algorithm for all $n \in \{a, \ldots, U\}$.

We are going to use Theorem 1 to find, for given δ, U and the lower bound a, the optimal parameter p. However, it is difficult to use Theorem 1 directly. We process as follows:

1. we find a good approximation $C^*(p, \delta, U)$ of the function $C_6(p, \delta, U)$,
2. for a given δ and U we find the probability $p^* = p^*(\delta, U)$ which minimizes the function $C^*(p, \delta, U)$,
3. we calculate the value $C_6(p^*, \delta, U)$.

Notice that this procedure does not yield the optimal probability, but only a reasonable one. However, numerical experiments shows that probabilities obtained are close to optimal ones. Note also that we do not consider the small factor D_6 during this process

We begin with the first component of $C_a(p, \delta, U)$. Namely, directly from the definition of the function B and from the equality $z\Gamma(z) = \Gamma(z+1)$, we get $B(n, 1+\alpha) = \frac{1}{n\prod_{j=1}^{n}(1+\frac{\alpha}{j})}$. Since $\prod_{j=1}^{\infty}(1 + (\frac{\alpha}{j})^2) = \frac{\sinh(\pi\alpha)}{\pi\alpha}$, we get $|nB(n, 1 + \alpha i)| \simeq \sqrt{\frac{\pi\alpha}{\sinh(\pi\alpha)}}$ for large n. Equation (1) shows that value $E[Y(n)]$ oscillates near the number $\frac{1-p}{p\log(\frac{1}{p})}$ for $n > 2$. Hence the function

$$\psi(p) = \frac{1-p}{p\log(\frac{1}{p})}\left(1 - 2\sqrt{\frac{2\pi^2}{\log(\frac{1}{p})\sinh(\frac{2\pi^2}{\log\frac{1}{p}})}}\right)$$

is an approximation of the lower bound on $E[Y(n)]$. In fact this approximation is very precise for all $p > \frac{1}{300}$ and $n \geq 6$. Therefore, if we consider probabilities $p \in [\frac{1}{300}, 1]$, then the function $\frac{1}{\psi(p)}$ is a precise upper bound of the number $\max_{6\leq m\leq U}\{\frac{1}{E[Y(m)]}\}$.

Next, from Equation (6) we deduce that the number $\frac{1}{2} + \frac{H_n}{\log(1/p)}$ is a very good approximation of the value $E[T(n)]$. Therefore we shall use the number $\frac{1}{2} + \frac{H_U}{\log(1/p)}$ as an upper bound of $E[T(n)]$.

Next, let $W(\delta, p, U)$ be defined by

$$\frac{1-p}{p\log(\frac{1}{p})} + \frac{2n(1-p)}{\log(\frac{1}{p})}\Re\left[\left(\frac{1}{p-\delta}\left(\frac{1-\delta}{p-\delta}\right)^{\frac{2\pi}{\log p}} - \frac{1}{p}\right)\sum_{k=1}^{4}B\left(n, 1 + \frac{2\pi ki}{\log p}\right)\right]$$

(see Equation 7). We finally put

$$C^*(p, \delta, U) = \frac{1}{\psi(p)}\left(\frac{1}{2} + \frac{H_U}{\log(1/p)} + W(\delta, p, U)\right).$$

and define

$$p^*(\delta, U) = \arg\min_{0\leq p\leq 1} C^*(p, \delta, U).$$

3.6 Comparisons with Previous Results and Experimental Data

The estimation of the probability p^* derived in the previous section as well as the estimation of the total expected number of slots are quite precise as confirmed by experimental data (see Table 1).

Up to now, we have considered the length of the slot as $\alpha \geq 1$. Since $\delta = \frac{\lambda}{\alpha-1}$, we have $\alpha = 1 + \frac{\lambda}{\delta}$. Hence the normalized time cost for initializing the network equals

$$\alpha \cdot (1 + C(p^*, \delta, U)) \cdot n = \left(1 + \frac{\lambda}{\delta}\right)(1 + C(p^*, \delta, U)) \cdot n .$$

Since we want to minimize the normalized time cost, we choose both parameters δ and p as follows:

$$(\delta^*, p^*) = \arg\min_{\delta, p}\left(1 + \frac{\lambda}{\delta}\right)(1 + C^*(p, \delta, U)) .$$

Table 1. Version 1: results for $\delta = 0.001$ and $n \geq 10$

U	$p^*(\delta, U)$	$(1 + \mathcal{C}(p^*, \delta, U)) \cdot n$	runtime of simulations
100	0.037678	$\leq 1.3271 \cdot n$	$\leq 1.3168 \cdot n$
1000	0.0267521	$\leq 1.3998 \cdot n$	$\leq 1.3398 \cdot n$
10000	0.0232507	$\leq 1.4677 \cdot n$	$\leq 1.3482 \cdot n$

Table 2. Version 2: results for $\lambda = 0.005$ and $n \geq 10$

U	p^*	δ^*	our algorithm	algorithm from [2]
100	0.0423848	0.0235993	$\leq 1.5927 \cdot n$	$\leq 1.6162 \cdot n$
1000	0.0267521	0.0221062	$\leq 1.6381 \cdot n$	$\leq 1.7497 \cdot n$
10000	0.0232507	0.0210236	$\leq 1.7647 \cdot n$	$\leq 1.9199 \cdot n$

Table 3. Version 2: results for $\lambda = 0.005$ and $n \geq 10$

U	p^*	δ^*	our algorithm
100	0.0423848	0.0235993	$\leq 1.4904 \cdot n$
1000	0.0267521	0.0221062	$\leq 1.5201 \cdot n$
10000	0.0232507	0.0210236	$\leq 1.5295 \cdot n$

Table 2 compares execution time of our algorithm with the algorithm from [2] for different parameter values.

Finally, let us note that it is possible to substitute lines 2-8 (Fig. 1) by a procedure developed in [3]. Then we have further improvements (see Table 3).

4   Conclusions and Future Works

We developed a simple initialization algorithm which utilizes carrier sensing. It outperforms the previous algorithms concerning the expected runtime. Moreover, we develop tools based on complex analysis and we provide a probabilistic analysis of algorithm's behavior. Previous work was based on simple estimations and experimental data only.

The presented method of finding almost optimal parameters of our algorithms is numerically complicated. One of our future goals is to simplify this process and investigate optimal parameters more carefully. Another point for further improvements is the optimization of the choice of the probability distribution for backoff values. This seems to be a challenging topic, but with fair chances for getting further substantial improvements. Similarly, one can use a tree-like structure to divide stations into groups in which a modified version of the algorithm from [2] is executed.

References

1. L. Ahlfors, *Complex Analysis: An Introduction to the Theory of Analytic Functions of One Complex Variable* 3rd edition, McGraw-Hill, 1979.
2. Z. Cai, M. Lu, X. Wang, *Distributed Initialization Algorithm for Single-Hop Ad Hoc Networks with Minislotted Carrier Sensing* IEEE, Transactions Parallel and Distributed Systems, vol. 14, no. 5, pp. 516-528, May 2003.
3. J. Cichoń, M. Kutyłowski, M. Zawada, *Initialization for Ad Hoc Radio Networks with Carrier Sensing and Collision Detection* 5th International Conference on AD-HOC Networks & Wireless 2006, LNCS 4104, Springer Verlag, Berlin 2006.
4. K. Jamieson, B. Hull, A. K. Miu, H. Balakrishnan, *Understanding the Real-World Performance of Carrier Sense* ACM SIGCOMM, Workshop on Experimental Approaches to Wireless Network Design and Analysis (E-WIND), Philadelphia, PA, August 2005.
5. Ph. Flajolet and R. Sedgewick, *Mellin transforms and asymptotics: Finite differences and Rice's integrals*, Theoretical Computer Science, vol. 144, no. 1&2, pp. 101-124, 1995.
6. I. S. Gradshteyn, I. M. Ryzhik, A. Jeffrey, *Table of Integrals, Series, and Products* 6th edition, Academic Press, 2000.
7. D. Knuth, *The Art of Computer Programming*, Addison-Wesley, 1997.
8. K. Nakano, S. Olariu, *Randomized Initialization Protocols for Ad Hoc Networks* IEEE, Transactions Parallel and Distributed Systems, vol. 11, no. 7, pp. 749-759, July 2000.
9. A. Micic, I. Stojmenovič, *A Hybrid Randomized Initialization Protocol for TDMA in Single-Hop Wireless Networks* Proceedings of the International Parallel and Distributed Processing Symposium, pp. 147-154, April 2002.
10. IEEE, *Wireless LAN Medium Access Control (MAC) and Physical Layer (PHY) specifications* IEEE 802.11 standards, June 1999.

Securing Communication Trees in Sensor Networks

Tassos D. Dimitriou

Athens Information Technology,
P.O. Box 68, 19.5 km Markopoulo Ave.,
GR-19002, Peania, Athens, Greece
tdim@ait.edu.gr

Abstract. In this work we present a protocol for securing communication trees in sensor networks. Communication trees can be used in a number of ways; to broadcast commands, to aggregate information or to route messages in the network. Broadcast trees are necessary for broadcasting information such as commands of maintenance packets from the base station to all sensors in the network. Aggregation trees are used in a complementary way to aggregate information collected by individual nodes so that meaningful summaries are presented to the base station, thus saving energy from unnecessary retransmissions. Alternatively, trees are constructed by many routing protocols, hence the security of these trees is important to prevent against many routing attacks.

In this work we demonstrate how to establish such trees in a secure and authenticated way. Our protocol is simple and efficient and furthermore enables changes to the tree structure due to failure of nodes or addition of new ones. Finally, it is resistant to a host of attacks that can be applied to sensor networks.

1 Introduction

Advances in MEMS and wireless technology have enabled the mass production of small computing devices equipped with sensing and communication capabilities. Such devices can be deployed in large numbers to form spontaneous networks capable of performing distributed sensing tasks. As sensor networks are usually deployed in environments with easy physical access, many of these applications require that data must be exchanged in a secure and authenticated manner. However, establishing secure communications between sensor nodes becomes a challenging task, given their limited processing power, storage, bandwidth and energy resources. Furthermore, existing protocols, including those developed for mobile ad-hoc networks do not seem to carry over in this domain.

Our work in this paper is inspired by the results of Karlof and Wagner [1] who demonstrate how vulnerable existing sensor network protocols are to simple and easily realizable attacks. In particular, we focus on securing communication trees that are used by many sensor network protocols as a tool for providing higher level services.

S. Nikoletseas and J.D.P. Rolim (Eds.): ALGOSENSORS 2006, LNCS 4240, pp. 47–58, 2006.

One such service is broadcasting [2,3,4]. Applications usually require broadcast operations to update global information or initiate network maintenance tasks such as updating topologies or executing route discoveries. Similarly, many applications periodically have to disseminate instructions to the network when there is need for sensing and collecting environmental data [5,6,7]. In most of these applications a tree needs to be constructed so that information reaches all the nodes without the problems introduced by flooding approaches.

While broadcasting is the process of disseminating information from a node to all other nodes in the network, aggregation or in-network processing is the reverse process, where data gathered by the sensors is processed within the network and only aggregated information is returned to the central location [7,8], thus reducing the volume of communicated data. Such use of in-network processing is used to achieve energy efficiency and increased lifetime of the network and has been applied to a number of routing protocols as well [9,10,11,12].

In this work we focus on securing such spanning-like communication trees by following a two step approach. First pairwise keys and other keys are established among neighboring nodes using a simple key establishment protocol. Then we show how to use the established keys to construct the tree in a secure and authenticated way that defends against the many attacks presented in [1].

Our protocol is simple and efficient since it does not involve the base station in establishing the security infrastructure. Furthermore, it offers resiliency against node capture and replication as security breaches remain localized. Finally, the protocol tolerates failures of existing nodes and allows for incremental addition of new ones in the tree.

2   Motivation

In this section we motivate our work by first presenting a simple protocol that attempts to build a routing tree rooted at the base station. This tree, for example, can be used to broadcast and authenticate commands, aggregate information, and so on. Based on this presentation we continue with a list of attacks that can be applied to this basic protocol.

The naive approach. Consider the following simple approach for constructing a spanning tree rooted at some base station. Initially all nodes are idle except from the base station which broadcasts a beacon message. This message may also contain some distance information from the base station if we want the spanning tree to correspond to a shortest path tree. In that case the beacon message has the form $(id, dist)$ and the base station transmits $(0, 0)$.

When a node receives a beacon for the first time, it records the unique identifier of the transmitting node, marks this node as its parent and updates its distance from the base station to $dist + 1$. At that moment the node becomes active and broadcasts to its neighbors the beacon containing the updated information. Any node that receives subsequent active beacons from any other node, compares the new beacon pair (id_{new}, d_{new}) with its current stored pair (id_s, d_s). If $d_{new} < d_s$, the node discards its old pair and stores the new one. Eventually, the outcome of this algorithm is a spanning tree that approximates a shortest path tree (the tree is not necessarily shortest as beacons may be delayed due to collisions and nodes may not be able to broadcast the new distances anymore). In any case, the tree is a spanning one as all reachable nodes will join the tree and this is the only requirement for our discussion.

Attacks. This simple protocol can be attacked in a number of ways [1]. For one thing any node can claim to be the base station thereby attracting all nodes to form a spanning tree with the attacker as the root. In such a case, all traffic will pass through the adversary who will have control of the network. Even if the attacker cannot initiate the protocol, it can still attract nodes by advertising its location with a strong signal: the adversary can use a powerful transmitter to broadcast a beacon loud enough to be heard by the entire network, causing every node to mark the adversary as its parent. Then it can drop packets at will, allowing a *selective forwarding* attack.

Other attacks include the *Sybil* attack, where the attacker by presenting multiple identities can appear to be in more than one places, hence attracting nodes to it, and the *wormhole* attack, where the adversary tunnels messages in one part of the network and replays them to some other part of it. The effect of this last attack is that nodes who normally would be multiple hops from a base station believe they are only one or two hops away via the wormhole. This can create a *sinkhole*: since the adversary on the other side of the wormhole can artificially provide a high-quality route to the base station, potentially all traffic in the surrounding area will be drawn to it.

Our goal in this work is to construct such spanning trees in a way that defends against both *outsider* and *insider* attacks. In outside attacks, the attacker has no special access to the sensor network, so encryption and authentication mechanisms seem to be the best choice against these types of attacks. However, more important from a security point of view is the existence of inside attacks. Now the attacker has compromised some node and is an authorized participant in the network, engaging in normal protocol operations.

3   The Protocol

In this section we present our protocol for building secure communication trees in sensor networks. The protocol consists of two phases: in the first phase nodes establish pairwise keys for communication with each other and the base station. In the second phase, the nodes use the pre-established keys to run the simple protocol presented in Section 2.

The intuition behind our protocol is simple. The discussion in Section 2 reveals that the protocol falls prey to the various attacks because nodes cannot really be sure who are their real neighbors. Many of the problems would be solved if sensors had a way to authenticate their neighbors and verify their *identities*. Another countermeasure would be to verify the *bi-directionality* of the links [1] since an attacker broadcasting with a powerful transmitter would not be able to hear the challenges of far away sensors and hence fail to respond. This would be a warning to the sensor node receiving a beacon message to exclude the attacker from its list of neighbors.

The goal of the key establishment phase is to enable the nodes verify the identities of their neighbors and establish common keys with them. Then beacon messages can be authenticated as they come from verified neighbors. The following notation will be used in the remaining of the paper:

Notation	Meaning
M_1, M_2	Concatenation of messages M_1 and M_2.
$E_K(M)$	Encryption of message M using key K.
$MAC_K(M)$	Message Authentication Code (MAC) of M using key K.
N_i	Nonce, random number used only once by node i.
K_{ij}	Symmetric key shared by nodes i and j

3.1   Key Establishment

The key establishment protocol enables the generation of three types of keys.

- A key K_i that can be used to secure communications of node i with the base station.
- A pairwise key K_{ij} that can be used to exchange routing or other information between node i and its neighbor j. This is a symmetric key, so both nodes need to agree on this common key.
- A group key K_i^G that allows node i to broadcast a message securely to all its neighbors. This allows a node to send an encrypted message to all its neighbors, thus avoiding unnecessary retransmissions.

K_i is not used in establishing the security infrastructure of the sensor network but only to encrypt data that must be seen by the base station. Hence it can be preloaded to the sensor nodes before deployment of the network. Alternatively, this key can be the result of the application of a secure one-way function F on a master secret K_m and the node ID i, i.e. $K_i = F(K_m, ID_i)$. We can think of this key as uniquely identifying each node. The one-wayness of F guarantees that if a node has been compromised and K_i has been known, an adversary cannot recover the master secret K_m or the keys of other nodes.

The other two keys can be established *on the fly* during network initialization. For that to happen we assume that all nodes come preloaded with a master secret K_m that is also known to the base station. The master secret will be used to secure information exchanged during the key establishment phase and to help derive the pairwise keys. Then it is *erased* from the memory of the sensor nodes.

Establishing pairwise keys. Pairwise keys are used to establish secure channels with a node's immediate neighbors and authenticate responses and messages coming from them. The pairwise key-establishment phase takes place during neighborhood discovery. Each node broadcasts a hello message containing its ID and a nonce to avoid replay attacks, properly MACed as follows:

$$Sensor_i \rightarrow Neighbors : ID_i, N_i, MAC_{K_m}(ID_i, N_i)$$

Upon receiving this message, sensor j uses the master key K_m to check the MAC. If the MAC verifies, node j adds node i in its list of *potential* neighbors. To verify the bi-directionality of the link node i must receive an acknowledgement by node j. Only then node j will become a valid neighbor to node i. To that respect, node j first computes the pairwise key K_{ij} as follows:

$$K_{ij} = F(K_m, i, j), \quad \text{where} \ i < j.$$

Then it responds with the following acknowledgement message:

$$Sensor_j \rightarrow Sensor_i : ID_j, N_j, MAC_{K_{ij}}(ID_i, ID_j, N_i + 1, N_j)$$

Observe that the previous message is MACed with the pairwise key K_{ij}, so node i first computes the pairwise key in a similar manner and then checks the MAC. If the MAC verifies, node j is added to the list of *valid* neighbors.

The use of the master key K_m is needed to establish the pairwise keys as nodes share no secret up to that point. Once this phase is over, K_m is *deleted* from the memory of the nodes. This ensures, that even if a node is compromised, the security breach remains *localized*. Hence the attacker will not be able to eavesdrop in other parts of the network or compromise the keys of other nodes.

An implicit assumption we make here is that the time required for the establishment of secure channels is smaller than the time needed by an adversary to compromise a node during deployment. As tamper resistance in sensor networks greatly increases the cost of sensor nodes, we must assume that an adversary needs more time to compromise a node and discover the master key K_m (see also [18] for a similar assumption). Preliminary experiments show that it takes about 100ms to establish pairwise keys with a node's neighbors, when the neighborhood size is 10, so this seems to be a realistic assumption.

Establishing group keys. Group keys are used so that messages can be propagated to *all* sensors within the neighborhood of a node i. One simple way to do this is to send a separate unicast message to each neighbor j encrypted and authenticated using the pairwise key shared between nodes i and j. While this is certainly one possibility, we feel that valuable energy is wasted as the same command must be transmitted unnecessarily many times.

A much simpler way for message propagation is for node i to transmit its group key K_i^G to its *valid* neighbors one by one, by encrypting it each time using the pairwise key K_{ij} it shares with neighbor j. Once the group key is settled, node i can broadcast commands to all valid neighbors using just one transmission. The message sent to sensor S_j is as follows:

$$Sensor_i : c = E_{K_{ij}}(\text{``Group key''}, K_i^G),$$
$$\tau = MAC_{K_{ij}}(ID_i, ID_j, c)$$
$$Sensor_i \rightarrow Sensor_j : ID_i, ID_j, c, \tau$$

First the node i encrypts the group key using an encryption key derived from K_{ij} and then creates a MAC τ of the resulting ciphertext c. Upon receipt of this message, node j verifies the MAC, retrieves the group key of node i and stores it along its pairwise key K_{ij} so that it can decrypt messages received by i.

At this point our key establishment phase is complete. Nodes have established pairwise and group keys, so we proceed in discussing the second phase of our protocol, the construction of the routing tree.

3.2   Building a Secure Communications Tree

The protocol for creating a breadth first tree is a straightforward extension of the one presented in Section 2. The base station (BS) initially "signs" the beacon pair $(ID, dist)$ using the group key K_{BS}^{G} it shares with its immediate neighbors (encryption is not really necessary here since the basic service we want to provide is integrity and authenticity of information). The message sent is

$$BS \rightarrow Neighbors : N_{BS}, (ID_{BS}, dist_{BS}), MAC_{K_{BS}^{G}}(N_{BS}, ID_{BS}, dist_{BS}).$$

When node i, a neighbor to the base station, receives the message for the first time, it marks the base station as its parent and updates its distance from the base station to $dist + 1$. At that moment the node becomes active and broadcasts to its neighbors a new beacon containing the updated information.

$$Sensor_i \rightarrow Neighbors : N, (ID_i, dist_i), MAC_{K_i^{G}}(N, ID_i, dist_i).$$

Any node that receives subsequent active beacons from any other node, compares the new beacon pair (id_{new}, d_{new}) with its current stored pair (id_s, d_s). If $d_{new} < d_s$, the node discards its old pair and stores the new one. As nodes can verify the origin of the beacons they will only accept transmissions coming from valid neighbors, hence this process of constructing the tree is a more secure one.

We feel however, that although our approach defends against outside attacks in which the adversary does not hold any keys, inside attacks are still possible after an adversary compromises a sensor node. In such a case, a malicious node may use the group key to send forged messages to neighboring nodes. To defend against such impersonation attack described above, we propose to use a *one-way hash key chain* computed by the base station. An one-way key chain is a sequence of keys, $k_0, k_1, \dots, k_{n-1}, k_n$ such that

$$\forall l, \quad 0 < l \leq n, \quad k_{l-1} = F(k_l),$$

where as usual F is a secure pseudo-random function that is difficult to invert. This means that having established k_j as authentic, one can easily verify that k_i, for $i > j$, is also authentic by hashing it $i - j$ times until one gets k_j. But having k_i one cannot come up with a k_l, for $l > i$ due to the one-wayness of F. Now, all is needed to use such a chain is for the base station to *commit* to the first key k_0. This key could for example be installed in the sensor nodes before deployment. Whenever the base station wishes to send a new message or command, it attaches to the message the *next* key from the hash chain.

In particular, the l-th beacon sent by BS contains the l-th commitment of the hash chain encrypted and authenticated using the base station's group key.

$$BS : c = E_{K_{BS}^G}(\text{"Beacon"}, k_l),$$
$$\tau = MAC_{K_{BS}^G}(c, ID_{BS})$$
$$BS \rightarrow Neighbors : ID_{BS}, c, \tau$$

A node receiving a command encrypted with the group key can verify its authenticity by checking whether the new commitment k_l generates the previous one through the application of F. When this is the case, it replaces the old commitment k_{l-1} with the new one in its memory and accepts the beacon as authentic. Then it marks the transmitting node as its parent and propagates the beacon and the key from the hash chain, this time encrypted and authenticated using its own group key. Hence beacon messages are authenticated as they reach deeper into the network. In all other cases, beacons are rejected.

Note, however, that although the possibility of an impersonation attack is reduced, it is not completely eliminated by this scheme. For example, an adversary may *jam* communications to sensor i so that it misses the last m commands and hence commitments. Then it introduces new commands by "recycling" the unused commitments.

One way to defend against this possibility is to assume that sensors are loosely synchronized and commands are issued only at *regular* time intervals. This is the approach taken in the μ-TESLA protocol [13]. In such a scenario, when the base station emits a beacon at the l-th time interval, it has to include the l-th key from the one way key chain. So, if a sensor does not receive anything within the next d time slots and the last commitment was k_l, it will expect to see the commitment k_{l+d} that accounts for the d missing messages. This way it cannot be fooled to authenticate unused commitments.

Another type of defense that not only eliminates but also helps detect the previous attacks is to have the sensor *acknowledge* the issued beacon using the key K_i it shares with the base station. In this manner, if the base station receives acknowledgements to beacons or messages that it has not issued, it will become aware that an attacker has compromised some node in the network. Of course this approach must be followed with caution. If all nodes respond to messages originated from the base station, traffic will increase disproportionately and hence valuable network energy would be wasted.

A simple twist to the previous proposal is to have sensors send an acknowledgement with some predefined *probability* p_{ack} that can be a function of node density and the likelihood of node compromise. This scheme would achieve a balance between increasing network traffic and detecting compromised nodes. In our continuous research we are planning to invest more on this simple approach and determine the probability of detecting an attack as a function of both the number of compromised nodes and p_{ack}.

Fault tolerance. Sensor nodes come with strict energy resources, so eventually nodes can fail due to power depletion. When this happens to a node, it may cause a number of problems depending on whether the node is a leaf or an internal node in the tree. If the sensor is a leaf then other nodes will eventually adjust their roles to fill the gap. If the sensor is an internal node, the communication tree is partitioned and nodes in the subtree rooted at the node become orphans.

Now consider such an orphan node u that is the root of the spanning tree sitting below it. This root needs to reconnect their orphaned subtrees to the main network. The way this happens is that each orphan u transmits to its local neighborhood a *search* message, indicating that it is looking for a parent. The format of this search message is shown below:

$$Sensor_u \rightarrow Neighbors : ID_u, N_u, MAC_{K_u^G}(\text{``Orphan''}, ID_u, N_u).$$

When this message is received by some prospective parent v, v responds with a message containing its ID and the distance from the base station, expressing its willingness to "adopt" the orphaned node. The message is MACed using the pairwise key shared by u and v.

$$Sensor_v \rightarrow Sensor_u : ID_v, MAC_{K_{uv}}(ID_v, dist_v, N_u + 1)$$

Upon receipt, the orphan then chooses the willing parent (perhaps using the distance metric) who must be a valid neighbor, i.e. one that was found in the key establishment phase.

Populating the network. In many circumstances there is need to refresh the network by adding new nodes to it. As these nodes do not have prior knowledge of the network topology they must establish pairwise keys with their neighbors and find their location in the communication tree. New sensor nodes come equipped with the master key K_m. Using K_m, a new node k can compute the private key and group key of an existing node i using the formulas[1]

$$K_i = F(K_m, i) \quad \text{and} \quad K_i^G = F(K_m, i, \text{``Group''}).$$

As we will see below, this knowledge is necessary so that it can convince existing nodes that it is a valid newcomer. In a similar manner it can compute the pairwise key K_{kj} it will share with some node j using the expression

$$K_{kj} = F(K_m, k, j), \quad \text{where} \quad k < j.$$

The next task is to find out about its neighbors and establish pairwise keys with them. First the new node broadcasts a node discovery request and then each node responds with its ID and distance pair $(ID_i, dist_i)$ *authenticated* using the key K_i. This is needed in order to prevent an adversary from realizing the following devastating attack. The adversary will send fake messages containing various IDs. When the new node makes the association between the ID and the pairwise key and stores it in its memory, the adversary can later compromise the node thus having acquired pairwise keys with every node in the neighborhood

---

[1] For simplicity we also assume that the group key can be generated by K_m.

(or even worse in the network). To prevent this type of impersonation attack the response sent by an existing node i is MACed as follows

$$ID_j, dist_j, N_j, MAC_{K_j}(ID_j, dist_j, N_j),$$

where N_j is a *fresh* nonce identifier, different from the ones used in the key establishment phase.

The new node, upon receiving these IDs, first computes K_j and then verifies the MAC. If everything is ok, it computes the pairwise key K_{kj} for each neighbor j and transmits it securely using the private key K_j of neighbor j. This is required since node j does not have the master key K_m anymore and cannot compute the pairwise key. The message sent is described below:

$$S_k : c = E_{K_j}(K_{kj}),$$
$$\tau = MAC_{K_j}(ID_k, ID_j, N_j + 1, c)$$
$$Sensor_k \rightarrow Sensor_j : ID_k, ID_j, c, \tau$$

Since K_j is known only to node j, node j accepts the new node as a valid neighbor. Furthermore, node k, indicates in the previous message which node will be its parent in the communication tree. When the phase is over, the master key is deleted from the memory of the new node. However, again we must assume that an adversary cannot compromise a new node during this phase. This is a valid assumption as the time to compromise a node and discover its keys is usually larger than the key establishment phase.

4   Security Analysis

We now analyze the security achieved by our scheme by discussing one by one some of the general attacks found in [1].

Spoofed, altered, replayed routing information. This attack targets information exchanged between nodes. However, this is not a problem in our case since information is encrypted and authenticated using the broadcast or pairwise keys.

Selective forwarding. Here an adversary selectively forwards certain packets while drops the rest. We feel that this attack is the most difficult to defend against as a compromised node will always have a chance to include itself in the flow of data. A standard defense against this type of attack is to allow the next hop to be chosen probabilistically, but in a tree setting this seems to be difficult. So, we propose here the use of multiple trees set up by the base station so that propagated information can be forwarded by multiple paths.

In general, to defend against compromised nodes requires security mechanisms with self-awareness and self-healing so that they can autonomously recognize and respond to security threats. This requires nodes to have knowledge about the network's state and be able to characterize normal and malicious behavior and take corrective actions. The challenge here is that all these steps must be taken autonomously, without human intervention [14].

Sinkhole and Wormhole attacks. In such attacks an adversary tunnels messages received in one part of the network and replays them in a different part, thus making this link attractive to nearby nodes and enabling selective forwarding.

In our protocol this attack can only take place during the key establishment phase or while adding new nodes in the network. But the authentication that takes place in those phases and its small duration makes this kind of attack difficult. Furthermore, our use of bi-directionality verification renders this attack useless. Finally, we should mention here another implicit defense when the tree is already constructed and distances to base station are known: when nodes see advertisements of high quality links contradicting their established distance to BS, they can set an alarm and warn the base station.

Sybil attacks. In a Sybil attack a node presents multiples identities, thus appearing to be in more than one place in the network. By doing so, an adversary may be able to control communications in different parts of the sensor field.

We feel that this attack does not apply in our case, since every node shares a unique symmetric key with the trusted base station and hence it cannot present multiple identities. An adversary may create clones of a compromised node and populate them into the node's neighborhood but this doesn't seem to offer any more advantages to the adversary than what she already has. Any attempts to populate different parts of the network will be rejected as the clones will not be able to convince existing nodes to authenticate and add them in the tree.

Hello flood attacks and Acknowledgment spoofing. Here the goal of the adversary is to attract nodes in the tree hierarchy to use an alleged high quality route by broadcasting with a powerful transmitter.

In our protocol, nodes broadcast or acknowledge HELLO messages during the neighborhood discovery phase. Since, however, messages are authenticated and bi-directionality is enforced, this attack is not possible. Notice here the use of bi-directionality verification. By insisting on the challenge-response mechanism, an attacker that is far away from a legitimate node will not be able to respond and hence it will not be considered a valid neighbor.

5   Related Work

Many applications in sensor networks (military, health, monitoring, etc.) require that data must be exchanged in a secure and authenticated way. However, establishing secure communications between sensor nodes becomes a challenging task, given their limited processing power, storage, bandwidth and energy resources.

Key establishment is required to bootstrap secure communications in sensor networks and there have been many protocols that attempt to establish pairwise keys between sensors. Their differences lie on the method of doing so. Random key pre-distribution schemes [16,17] deploy nodes with a predefined list of keys. Then nodes establish connections with their neighbors if they have at least one key in common with them. These schemes offer, however, only "probabilistic" security as compromise of a single node may result in a breach of security in some other part of the network. Furthermore, these schemes seem to be more vulnerable to attacks applied in the initial, key establishment phase.

Other protocols [18,19,20] follow a master key approach where the keys are created during network initialization. These methods are preferable in our opinion, because even if a node is compromised, it cannot be used to populate other parts of the network and security breaches remain localized. However, none of these methods was created for securing communication trees in sensor networks.

A proposal for authenticated broadcast is the μ-TESLA protocol [13]. TESLA achieves authenticated broadcast by using delayed key disclosure and one-way key chains. Replay is prevented because messages authenticated with previously disclosed keys are ignored. TESLA, however, requires loose time synchronization and cannot be used to secure communication trees.

6   Conclusions

In this work we have presented a protocol that attempts to secure the construction of communication trees in sensor networks. These trees can be used as building blocks of many higher level services and come into many disguises; as broadcast, as aggregation or as routing trees.

Our protocol consists of two phases: in the first, pairwise and other keys are established among neighboring nodes, while in the second the actual tree is built up. Both phases are very simple and involve a minimal number of message exchanges. In both phases we focus on neighbor and bi-directional link verification, that seems to be important in defending against many attacks in sensor network protocols. We have also demonstrated how to add new nodes to the tree and how to tolerate node failures. Finally, the proposed protocol scales very efficiently as the key establishment mechanisms are very efficient in terms of computation, communication and storage requirements.

Our work is the first that attempts to secure such trees and we hope that it will initiate further research into this area.

References

1. C. Karlof and D. Wagner, "Secure routing in wireless sensor networks: Attacks and countermeasures," *Elsevier's Ad Hoc Network Journal, Special Issue on Sensor Network Applications and Protocols*, vol. 1, pp. 293–315, September 2003.
2. S.Y. Ni, Y.C. Tseng, Y.S. Chen, and Sheu J.P. "The broadcast storm problem in a mobile ad hoc network," *In Proceedings of the ACM/IEEE International Conference on Mobile Computing and Networking (MOBICOM)*, pages 151162, 1999.
3. J.E. Wieselthier, G.D. Nguyen, and A. Ephremides, "On the construction of energy-efficient broadcast and multicast trees in wireless networks," in *Proceedings of IEEE INFOCOM*, pages 589594, Apr. 2000.
4. I. Stojmenovic, M. Seddigh, and J.D. Zunic, "Dominating sets and neighbor elimination-based broadcasting algorithms in wireless networks," In *IEEE Trans. Parallel Distrib. Syst.*, 2002.

5. M. Chang, and N. Liu, "Optimal controlled flooding search in a large wireless network," In *Proc. 3rd Inter. Symposium on Modeling and Optimization in Mobile, Ad Hoc and Wireless Networks*, 2005.
6. N. Sadagopan, B. Krishnamachari, and A. Helmy, "The ACQUIRE mechanism for efficient querying in sensor networks," In *IEEE International Workshop on Sensor Network Protocols and Applications (SNPA03)*, May 2003.
7. C. Intanagonwiwat, R. Govindan, D. Estrin, J. Heidemann, and F. Silva, "Directed diffusion for wireless sensor networking," *ACM/IEEE Transactions on Networking*, vol. 11, February 2002.
8. B. Krishnamachari, D. Estrin, S. Wicker, "Modeling Data Centric Routing in Wireless Sensor Networks," in *Proceedings of IEEE INFOCOM*, New York, NY, June 2002.
9. Y. Yao and J. Gehrke, "The cougar approach to in-network query processing in sensor networks," in *SIGMOD Record*, September 2002.
10. W. Heinzelman, J. Kulik, and H. Balakrishnan, "Adaptive protocols for information dissemination in wireless sensor networks," in *Proceedings of the 5th Annual ACM/IEEE International Conference on Mobile Computing and Networking (MobiCom99)*, Seattle, WA, August 1999.
11. C. Schurgers and M.B. Srivastava, "Energy efficient routing in wireless sensor networks," in *MILCOM Proceedings on Communications for Network-Centric Operations: Creating the Information Force*, McLean, VA, 2001.
12. Kemal Akkaya and Mohamed Younis, "A Survey on Routing Protocols for Wireless Sensor Networks", Journal of Ad Hoc Networks.
13. A. Perrig, R. Szewczyk, V. Wen, D. Culler, and J. Tygar, "SPINS: Security Protocols for Sensor Networks," In *Proc. of 7th Annual ACM Mobicom*, 2001.
14. Tassos Dimitriou, I. Krontiris, "Autonomic Communication Security in Sensor Networks," In *2nd International Workshop on Autonomic Communication*, WAC 2005.
15. D. Carman, P. Kruus, and B.J.Matt, "Constraints and approaches for distributed sensor network security," Tech. Rep. 00-010, NAI Labs, June 2000.
16. L. Eschenauer and V. D. Gligor, "A key-management scheme for distributed sensor networks," in *Proceedings of the 9th ACM conference on Computer and communications security*, pp. 41–47, 2002.
17. H.Chan, A.Perrig, and D.Song, "Random key predistribution schemes for sensor networks," in *IEEE Symposium on Security and Privacy*, pp. 197–213, May 2003.
18. S. Zhu, S. Setia, and S. Jajodia, "LEAP: efficient security mechanisms for large-scale distributed sensor networks," in *Proceedings of the 10th ACM CCS*, pp. 62–72, October 2003.
19. T. Dimitriou, and I. Krontiris, "A Localized, Distributed Protocol for Secure Information Exchange in Sensor Networks," in *Proc. of the 5th IEEE Inter. Workshop on Algorithms for Wireless, Mobile, Ad Hoc and Sensor Networks*, 2005.
20. Tassos Dimitriou, "Efficient Mechanisms for Secure Inter-node and Aggregation Processing in Sensor Networks", in *4th International Conference on Adhoc Networks and Wireless-ADHOC NOW*, 2005

Self-deployment Algorithms for Mobile Sensors on a Ring

Paola Flocchini[1], Giuseppe Prencipe[2], and Nicola Santoro[3]

[1] School of Information Technology and Engineering
University of Ottawa, Canada
flocchin@site.uottawa.ca
[2] Dipartimento di Informatica
Università di Pisa, Italy
prencipe@di.unipi.it
[3] School of Computer Science
Carleton University, Ottawa, Canada
santoro@scs.carleton.ca

Abstract. We consider the *self-deployment* problem in a ring for a network of identical sensors: starting from some initial random placement in the ring, the sensors in the network must move, in a purely decentralized and distributed fashion, so to reach in finite time a state of static equilibrium in which they evenly cover the ring. A self-deployment algorithm is *exact* if within finite time the sensors reach a static *uniform* configuration: the distance between any two consecutive sensors along the ring is the same, d; the self-deployment algorithm is ϵ-*approximate* if the distance between two consecutive sensors is between $d - \epsilon$ and $d + \epsilon$.

We prove that *exact* self-deployment is *impossible* if the sensors do not share a common orientation of the ring.

We then consider the problem in an oriented ring. We prove that if the sensors know the desired final distance d, then *exact* self-deployment is possible. Otherwise, we present another protocol based on a very simple strategy and prove that it is ϵ-approximate for any chosen $\epsilon > 0$.

Our results show that a *shared orientation* of the ring is an important computational and complexity factor for a network of mobile sensors operating in a ring.

1 Introduction

1.1 The Framework

A mobile sensors network is composed of a distributed collection of sensors that in addition to the traditional sensing, computation, and communication capabilities of static sensors, have also locomotion capabilities. Mobility facilitates a number of useful network capabilities; for example, they can patrol a wide area, they can be re-positioned for better surveillance, etc.; moreover, they are especially useful in environments that may be both hostile and dynamic. There have been some research efforts on the deploying of mobile sensors, most of them

S. Nikoletseas and J.D.P. Rolim (Eds.): ALGOSENSORS 2006, LNCS 4240, pp. 59–70, 2006.

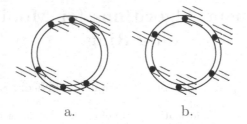

Fig. 1. Starting from an initial arbitrary placement (a), the sensors must move to a uniform cover of the ring (b)

based on centralized approaches; e.g., [22] assumes that a powerful cluster head is available to collect the sensor location and determine the target location of the mobile sensors.

Locomotion however allows the sensors to *self-deploy*; that is, starting from some initial random configuration, the sensors in the network can spread out in a purely decentralized and distributed fashion, and cover the area satisfying some optimization criteria (e.g., evenly, maximizing coverage, etc.) [11,12,13,15,21]. In contrast to [12] where the sensors are deployed one at the time, we consider the case when the sensors are deployed at the same time and they organize themselves in an adaptive manner. Unlike [15], we do not require prespecified destinations for the sensors, and unlike [12] we do not assume the sensors know where they are, since for small sensors localization is very hard. An essential requirement is that the network will reach a state of static equilibrium within finite time.

The *self-deployment* problem is quite similar to the *scattering* or *coverage* problem considered in cooperative mobile sensorics (e.g., [1]), and related to the *formation* problem (e.g. [10,18,19]); a key difference in these investigations is that usually there is no requirement that the network reaches a state of static equilibrium.

1.2   The Problem

In this paper, we are interested in the self-deployment of a mobile sensor network in a *ring* (e.g., a circular rim, as shown in Figure 1): starting from an initial random placement on the ring, the sensors must within finite time position themselves along the ring at (approximately) equal distance. A self-deployment algorithm, the same for all sensors, will specify which sequence of operations (communication/sensing, computing a destination, moving towards a point) a sensor must perform whenever it is active. We say that a self-deployment algorithm is *exact* if within finite time the sensors reach a *uniform* configuration: the distance between any two consecutive sensors along the ring is the same, d. We say that a self-deployment algorithm is ϵ-*approximate* if the distance between two consecutive sensors is between $d - \epsilon$ and $d + \epsilon$.

A self-deployment algorithm has recently been developed for the *line* [5] (e.g., a rectilinear corridor), and one has been designed for the *ring* as part of a larger protocol for uniform circle formation [2,6,14,17,20]. Both protocols yield only *approximate* solutions. However, they operate even with very weak sensors: anonymous (i.e., the sensors are indistinguishable), oblivious (i.e., each sensor has no memory of past actions and computations), asynchronous (i.e., each sensor becomes active at unpredictable times and the duration of its actions is unpredictable), and without a common coordinate system (e.g., no access to GPS). To date, no *exact* solution exists for these types of sensors.

1.3   Our Results

We first prove a strong negative result. In fact, we prove that *exact* self-deployment is actually *impossible* if the sensors do not share a *common orientation* of the ring; notice that this is much less a requirement than having global coordinates or sharing a common coordinate system. This impossibility result holds even if the sensors have unlimited memory and unbounded computational power, and even if all their actions, when active, are instantaneous and their visibility/communication radius is unlimited.

Faced with this strong negative result, the interesting question becomes under what restriction the self-deployment problem can be solved with an exact algorithm. Since the impossibility result holds in absence of common orientation of the ring, we consider the problem in *oriented* rings.

We prove that, in an oriented ring, if the sensors know the desired final distance d, then *exact* self-deployment is possible. In fact we present a simple protocol and prove that it allows the sensors to deploy themselves uniformly along the ring in finite time. This positive result holds even for the weakest sensors: anonymous, oblivious, asynchronous, with no common coordinate system; it works correctly even when every sensor can "locate" only its two neighbors or when the sensors have only a fixed sensing radius $v > d$.

Finally we turn to the case of an oriented ring when the desired final distance d is unknown. We present another protocol based on a very simple strategy and prove that it is ϵ-approximate for any fixed $\epsilon > 0$. As in [3,4,5], the difficulty is not in the protocol but in the proof of its correctness. Also in this case, the protocol works even for the weakest sensors: anonymous, oblivious, asynchronous, with no common coordinate system. The algorithm works correctly even when every sensor can "locate" only its two neighbors or when the sensors have only a fixed sensing radius $v \geq 2d$.

In the last protocol, the strategy we use is *go-to-half*. Interestingly was shown by Dijkstra [7] that in the *unoriented* ring *go-to-half* does *not* converge, and hence can not be used for self-deployment[1]. In other words, as already shown by our impossibility result, our result stresses that a *shared ring orientation* is an important computational and complexity factor for a network of mobile sensors

---

[1] It does however converge in a *line* as recently shown by Peleg [5] with a very involved proof.

operating in a ring. For space constraints, some of the proofs will be omitted and can be found in [8].

1.4 Related Work

The self-deployment problem has been investigated with the goal to cover the area so to satisfy some optimization criteria (e.g., evenly, maximizing coverage, etc.) [11,12,13,15,21]. For example, in [21] the problem is to maximize the sensor coverage of the target area minimizing the time needed to cover the area. Typically, distributed self-deployment protocols first discover the existence of coverage holes (the area not covered by any sensor) in the target area based on the sensing service required by the application. After discovering a coverage hole, the protocols calculate the target positions of these sensors, that is the positions where they should move. Loo et al. [15] considered a system consisting of a number of cooperating mobile nodes that move toward a set of prioritized destinations under sensing and communication constraints; unlike them, we do not require prespecified destinations for the sensors. Howard et al. [12] address the problem of incremental deployment, where sensors are deployed one-at-a-time into an unknown environment, and each sensor uses information gathered by previously deployed sensors to determine its deployment location. They assume every sensor is equipped with an ideal localization sensor. We do not assume the sensors know where they are, since for small sensors localization is very hard. The goal is to maximize network coverage under the constraint that nodes maintain line-of-sight with each other.

The self-deployment problem is related to a well studied problem in the field of autonomous mobile sensors: that of the *pattern formation* [9,10,18,19]; in particular to the one of *uniform circle formation* [2,6,17]. In this problem, very simple sensors are required to uniformly place themselves on the circumference of a circle not determined in advance (i.e., the sensors do not know the location of the circle to form). The main difference between these robotics investigations and our self-deployment problem in the ring is that in those problems, the sensors can freely move on a two dimensional plane; in contrast, our sensors can move only on the ring.

The strategy *go-to-half*, that we employ in one of our protocols was first analyzed by Dijkstra [7]; he showed that in an *unoriented* ring *go-to-half* does *not* converge (and hence can not be used for self-deployment). Recently, *go-to-half* has been shown by Peleg [5] (with a very involved proof) to converge in a *line*. Convergence in the unoriented ring has been announced for the *go-to-half-half* strategy by Défago and Konagaya [6,17].

2 Terminology and Model

We consider a sensors network in a ring (i.e., a circular line). Let $s_1, \ldots, s_n$ be the n sensors initially randomly placed on the ring (Figure 1). Let $d_i(t)$ be the distance between sensor s_i and sensor s_{i+1} at time t. When no ambiguity arises, we will omit the time and simply indicate the distance as d_i.

We will use a very general definition of a sensor as a computational unit capable of sensing (e.g., by communication) the positions of other sensors in its surrounding (within a fixed radius), performing local computations on the located/communicated data, and moving towards the computed destination. The local computation is done according to a deterministic algorithm that takes in input the located/communicated data, and returns a destination point towards which the sensor moves. All the sensors execute the same algorithm.

Each sensor repeatedly cycles through four *states*: when active, a sensor determines the positions of the other sensors in its radius – *Locate*; it computes the next destination point by executing the algorithm – *Compute*; and it moves towards the computed point – *Move*; after such a move the sensor may become inactive – *Wait*. The sequence: *Wait - Locate - Compute - Move* form a *computation cycle* (or briefly *cycle*) of a sensor. In the following, the "view of the world" of a sensor is defined as a snapshot of the positions of the other sensors in its own coordinate system (obtained in the *Locate* state).

The sensors are completely *autonomous*: no central control is needed. Furthermore they are *anonymous*, meaning that they are a priori indistinguishable by their appearance, and they do not (need to) have any kind of identifiers that can be used during the computation. They are *oblivious*: each sensor has no memory of past actions and computations; in other words, the computation is based solely on what located in the current cycle.

In general, no assumptions on the cycle time of each sensor and on the time each sensor takes to execute each state of a given cycle are made. It is only assumed that each cycle is completed in finite time, and that the distance traveled in a cycle is finite. Moreover, the sensors do not need to have a common notion of time, and each sensor can execute its actions at unpredictable time instants: this scenario is called *asynchronous* (ASYNC).

We also consider (in our impossibility result) a different scenario, where there is a global clock tick reaching all sensors simultaneously, and a sensor's cycle is an instantaneous event that starts at a clock tick and ends by the next. This scenario is called *semi-synchronous* (SSYNC). The only unpredictability is given by the fact that at each clock tick, every sensor is either *active* or *inactive*, and only active sensors perform their cycle. The unpredictability is restricted by the fact that at least one sensor is active at every time instant, and every sensor becomes active at infinitely many unpredictable time instants.

Let us denote by AS and SS the class of problems that are solvable in the asynchronous and in the semi-synchronous setting, respectively. Then,

Theorem 1 ([16]). AS ⊂ SS.

3 Impossibility of Exact Self-deployment

In this section, we show that the exact self-deployment problem is unsolvable; in other words, given a set of sensors placed on the rim of a circle, there exists no deterministic algorithm that, in a finite number of cycles, places the sensors uniformly on the ring.

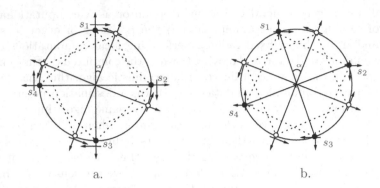

a.   b.

Fig. 2. (a) An example of starting configuration for the proof of Theorem 2. The black sensors are in S_1, while the white ones in S_2. (b) Theorem 2: the adversary moves only sensors in S_1.

Theorem 2. *Let $s_1, \ldots, s_n$ be all on a ring C. Then, in SSYNC, there is no deterministic exact self-deployment algorithm.*

Proof. By contradiction, let us assume there exists a deterministic algorithm $\mathcal{A}$ that solves the problem in a finite number of cycles. Furthermore, let us assume that there is an even number of sensors placed on C, and that the n sensors can be split in two subsets according to their views of the world. In particular, in the first subset, call it S_1, there are $s_1, \ldots, s_{n/2}$, and in the second subset, call it S_2 the other sensors. The sensors in S_1 and S_2 are placed on the vertices of two regular $n/2$-gons, and the two polygons are rotated of an angle smaller than $360°/n$. Furthermore, all sensors have their local coordinate axes rotated so that they all have the same view of the world (refer to Figure 2.a for an example).

Lemma 1. *If activating only the sensors in S_1 no exact self-deployment on C is reached, then also activating only the ones in S_2 no exact self-deployment on C is reached.*

Lemma 2. *If activating only the sensors in S_1 an exact self-deployment on C is reached, then also activating only the sensors in S_2 an exact self-deployment on C is reached. Moreover, activating both sets no exact self-deployment on C is reached.*

In the following, we define an adversary so that $\mathcal{A}$ never succeed in solving the problem. Algorithm 1 reports the protocol followed by the adversary.

First we note that, by the way the adversary is defined and since the sensors in S_1 (resp. S_2) have the same view, these sensors will always move together (when all activated). In the following, we will prove by induction the following property $\mathcal{P}rop$:

> for all $t \geq 0$, the sensors all have the same view of the world and are not in an exact self-deployment on C.

Algorithm 1. The Adversary

(a) If activating only the sensors in S_1 no exact self-deployment on $\mathcal{C}$ is reached. Activates all sensors in S_1, while all sensors in S_2 are inactive, and goto (c). Otherwise,
  (b) If activating only the sensors in S_2, no exact self-deployment on $\mathcal{C}$ is reached. In this case, it activates all sensors in S_2, while all sensors in S_1 are inactive, and goto (c). Otherwise, all sensors are activated, and goto (c).
(c) If activating only the sensors in S_2 no exact self-deployment on $\mathcal{C}$ is reached. In this case, it activates all sensors in S_2, while all sensors in S_1 are inactive, and goto (a). Otherwise,
  (d) If activating only the sensors in S_1 no exact self-deployment on $\mathcal{C}$ is reached. In this case, it activates all sensors in S_1, while all sensors in S_2 are inactive, and goto (a). Otherwise, all sensors are activated, and goto (a).

By construction, $\mathcal{P}rop$ is clearly true at $t = 0$. Let us assume it is true at a given time $t > 0$. We distinguish the possible cases.

1. If the check performed in (a) is true, then clearly at time $t + 1$ there is no exact self-deployment on $\mathcal{C}$. Furthermore, all sensors will still have the same view of the world (see the example depicted in Figure 2.b).
2. If the check performed in (a) is true, then rule (b) is executed. Two subcases can occur.
   3.1. If the check of rule (b) is false, then at time $t + 1$ there is no exact self-deployment on $\mathcal{C}$, and all sensors have the same view of the world.
   3.2. Otherwise, all sensors are activated at time t, and by Lemma 2 no exact self-deployment on $\mathcal{C}$ is reached at time $t + 1$.
3. Rules (c) and (d) are handled symmetrically to previous rules (a) and (b).

Therefore, there is no time $t' \geq t$ so that the sensors are in a exact self-deployment on $\mathcal{C}$, having a contradiction.

By Theorem 1, we have

Corollary 1. *Let $s_1, \ldots, s_n$ be all on $\mathcal{C}$. Then, in* ASYNC *there is no deterministic algorithm that brings them uniformly distributed on $\mathcal{C}$ in a finite number of cycles.*

4   Self-deployment in an Oriented Ring: Interdistance Known

In this section we assume that the final distance d between two sensors is known to them. Moreover, the sensors have a fixed visibility radius of $2d$ and they can only locate up to such distance.

4.1   The Algorithm

The algorithm is very simple: sensors asynchronously and independently observe clockwise at distance $2d$, then they position themselves at distance d from the closest observed sensor (if any).

Protocol UNIFORM KNOWN (for sensor s_i)

- Locate clockwise at distance $2d$. Let d_i be the distance to next sensor. If none, $d_i = 2d$.
- If $d_i \leq d$ do not move.
- If $d_i > d$ move clockwise and place yourself at distance d from s_{i+1}.

4.2   Correctness

We say that a sensor is *white* if its distance to the clockwise neighbor is greater than or equal to d. We say that a sensor is *gray* if such a distance is smaller than d. Moreover we say that a white sensor is *good* if its distance to the clockwise neighbor is exactly d, it is *large* if its distance is strictly greater than d.

We call a *white bubble* a sequence of consecutive white sensors delimited by grey sensors. Let $W = s_i, s_{i+1}, \ldots, s_{i+m}$ be a white bubble. Sensor s_{i-1} is said to be the predecessor of the bubble, sensor s_{i+m+1} is the successor. Clearly predecessors and successors of a white bubble are gray, unless the ring contains white sensors only; notice that in this case all sensors are good. The size of W, indicated as $|W|$ is the number of white sensors composing the bubble (in this example m), its length, indicated by $l(W)$, is the length of the ring between the predecessor of the white bubble and its successor (assuming not all sensors are white); i.e., $l(W) = \sum_{j=-1}^{m} d_{i+j}$. Similarly, we define a *gray bubble* $G = s_i, s_{i+1}, \ldots, s_{i+m}$ as a sequence of consecutive gray sensors delimited by white sensors. Its size $|G|$ is the number of gray sensors in G; the length $l(G)$ is defined as the length of the ring between the first and the last gray sensor in G (note that this definition is different from $l(W)$).

The next two lemmas contain some simple facts.

Lemma 3. *At each point in time, if there are gray sensors, then the number of white bubbles equals the number of gray bubbles.*

Lemma 4. *At each point in time, if there are grey sensors there must be at least a bubble (i.e., a large sensor).*

Lemma 5. *A white sensor cannot become gray.*

Proof. In order for a white sensor s_j to become gray, its distance to the next sensor s_{j+1} should become smaller than d. By definition, sensors move clockwise and move according to the algorithm; so sensor s_{j+1} will never get closer to s_j. On the other hand, by definition of our algorithm, sensor s_j will never move at a distance smaller than d to s_{j+1}.

Lemma 6. *Let $W = s_i, s_{i+1}, \ldots s_{i+m}$ be a white bubble in the ring at time t. If $l(W) \geq d \cdot (|W|+1)$, in finite time, say at time t', the size of the bubble increases.*

Lemma 7. *Let $W_1, \ldots W_z$ be the white bubbles present in the ring at time t. At least one of these bubble W_k is such that $l(W_k) \geq d \cdot |W_k| + 1$.*

By Lemmas 6 and 7, we have that:

Lemma 8. *The number of grey sensors decreases.*

Finally, by Lemmas 5 and 8 we derive the main theorem.

Theorem 3. *In finite time all sensors are good.*

5 Self-deployment in an Oriented Ring: Interdistance Unknown

In this section we assume that each sensor has a fixed visibility radius of v, and does not know the final interdistance d between the sensors. Although d is not known, we must have that $v > 2d$ for our algorithm to work.

5.1 The Algorithm

Also this algorithm is very simple: sensors asynchronously and independently locate in both directions at distance v, then they position themselves in the middle between the closest observed sensor (if any).

Protocol UNIFORM UNKNOWN (for sensor s_i)

- Locate around at distance v. Let d_i be the distance to next sensor, d_{i-1} the distance to the previous (if no sensor is visible clockwise, $d_i = v$, analogously for counterclockwise).
- If $d_i \leq d_{i-1}$ do not move.
- If $d_i > d_{i-1}$ move to $\frac{d_i + d_{i-1}}{2} - d_{i-1}$ clockwise.

5.2 Correctness

Let $d_{min}(t) = Min\{d_i(t)\}$ and $d_{max}(t) = Max\{d_i(t)\}$. Let C be the length of the circumference of the ring. First observe the following simple fact:

Lemma 9. *We have that: $\forall t$, $d_{min}(t) \leq d$ and $d_{max}(t) \geq d$.*

Proof. By contradiction. Let the minimum distance be greater than d. We would have that $C > k \cdot d$, which is impossible since by definition $C = k \cdot d$. Same argument holds for d_{max}.

The next lemma shows that if, at some point there is a unique minimum (resp. maximum) interval, it will become bigger (resp. smaller).

Lemma 10. *If at time t there is a unique minimum interval, we have that:*
$\forall t, \exists t' > t : d_{min}(t') > d_{min}(t)$. *If at time t there is a unique maximum interval, we have that:* $\forall t, \exists t' > t : d_{max}(t') < d_{max}(t)$.

Proof. Let s_{j-1} and s_j be the sensors that delimit the minimum interval $[s_{j-1}, s_j]$, whose length is $d_{j-1}(t) = d_{min}(t)$ at time t. First observe that, since $d_{j-2}(t) > d_{j-1}(t)$, by the algorithm we know that sensor s_{j-1} does not move at time t; actually, it will not be able to move as long as d_{j-2} remains greater than d_{j-1} (i.e., as long as s_j does not move). Consider now the first time t' when s_j is activated. Since s_{j-1} has not moved from time t to time t', we have that, at time t', $d_{j-2}(t')$ is still greater than $d_{j-1}(t')$. At time t', s_i then moves following the rule of the algorithm and $d_{j-1}(t') = \frac{d_{j-1}(t)+d_j(t')}{2} \geq \frac{d_{j-1}(t)+d_j(t)}{2} > d_{j-1}(t)$. Similar argument holds for d_{max}.

We now show that if at some point there are several minimum (resp. maximum) intervals of a certain length, their number will decrease.

Lemma 11. *If at time t there are $r > 1$ minimum intervals of length $d_{min}(t)$, either all intervals have length d and the sensors are deployed, or there exists a time $t' > t$ when the number of minimum intervals of length $d_{min}(t)$ is $r' < r$.*

Analogously,

Lemma 12. *If at time t there are $r > 1$ maximum intervals, either all intervals have length d and the sensors are deployed, or there exists a time t' when the number of maximum intervals is $r' < r$.*

We now show that the minimum intervals converge to a value $A = d - \gamma_{min}$, with $\gamma_{min} \geq 0$, and the maximum intervals converge to a value $B = d + \gamma_{min}$, with $\gamma_{max} \geq 0$.

Lemma 13. *Let $d_{min}(t)$ (resp $d_{max}(t)$) be the distance of a minimum (resp. maximum) interval at time t. We have that, for any arbitrary small $\epsilon > 0$ there exists a time $t' > t$ such that, $\forall t'' > t'$: $|d_{min}(t'') - A| \leq \epsilon$, and, $\forall t'' > t'$: $|d_{max}(t'') - B| \leq \epsilon$.*

Proof. From Lemmas 10 and 11 the intervals must converge; from Lemma 9 the minimum must converge to a value smaller than (or equal to) d, and the maximum must converge to a value greater than (or equal to) d.

Let us call A-*regular* at time t an interval that, at time t is ϵ-close to A; that is an interval whose length $d_j(t)$ is such that $|d_j(t) - A| \leq \epsilon$. Analogously, we call B-*regular* an interval that is ϵ-close to B. We call A-*irregular* at time t an interval that, at time t, is smaller than d, but not ϵ-close to A; B-*irregular* one that is greater than d, but not ϵ-close to B.

The following lemma shows that there exists a time t, after the time when the previous Lemma 13 holds, when any interval greater than the minimum (and smaller than d) is A-*regular*, and any interval smaller than the maximum (and greater than d) is *B-regular*. In other words, each interval is either ϵ-close to A or to B. Notice that this property is not obvious; in fact, the only thing we know up to now is the convergence to A and B of the minimum/maximum intervals over time, while nothing is known about the other intervals.

Lemma 14. *Let $\epsilon > 0$ be arbitrarily small, and let t'_ϵ be a time when Lemma 13 holds. There exists a time $t''_\epsilon > t'_\epsilon$ when: for all intervals $[s_j, s_{j+1}]$ with $d_j(t'') \leq d$, $|d_j(t''_\epsilon) - A| \leq \epsilon$; for all intervals $[s_i, s_{i+1}]$ with $d_i(t''_\epsilon) \geq d$, $|d_i(t'') - B| \leq \epsilon$.*

Lemma 15. *Let t be a time when Lemma 14 holds. If at some time $t' > t$ at least an interval becomes irregular, then there exists a time $t'' > t'$ when all intervals are irregular.*

We now show that, after a time when Lemma 14 holds, all intervals actually converge to d (i.e., $A = B = d$).

Lemma 16. *Let $\epsilon > 0$ be arbitrarily small, and let t'_ϵ be a time when Lemma 14 holds. If $B - A > 2\epsilon$, at least an interval becomes irregular.*

Theorem 4. *For any arbitrary small $\epsilon > 0$ there exists a time t, such that $\forall t' > t, \forall i: |d_i(t') - d| \leq \epsilon$.*

Proof. By contradiction. Let $A \neq B$. From Lemma 14, there is a time t when all intervals are ϵ-close to A and B. From Lemma 16, at least one interval will become irregular at some time $t' > t$. However, by Lemma 15 there is a time $t'' > t'$ when all intervals become irregular (including the minimum and the maximum). This contradicts Lemma 13.

Acknowledgments. The authors would like to thank Vincenzo Gervasi, Toni Mesa, and Linda Pagli for the many discussions and suggestions.

References

1. Y. U. Cao, A. S. Fukunaga, A. B. Kahng, and F. Meng. Cooperative Mobile Robotics: Antecedents and Directions. In *IEEE/TSJ International Conference on Intelligent Robots and Systems*, pages 226–234, 1995. Yokohama, Japan.
2. I. Chatzigiannakis, M. Markou, and S. Nikoletseas. Distributed Circle Formation for Anonymous Oblivious Robots. In *Experimental and Efficient Algorithms: Third International Workshop (WEA 2004)*, volume LNCS 3059, pages 159 –174, 2004.
3. R. Cohen and D. Peleg. Convergence Properties of the Gravitational Algorithm in Asynchronous Robot Systems. In *Proc. of the 12th European Symposium on Algorithms*, volume LNCS 3221, pages 228–239, 2004.
4. R. Cohen and D. Peleg. Robot Convergence via Center-of-Gravity Algorithms. In *Proc. of the 11th Int. Colloquium on Structural Information and Communication Complexity*, volume LNCS 3104, pages 79–88, 2004.
5. R. Cohen and D. Peleg. Local Algorithms for Autonomous Robot Systems. In *Proc. of the 13th Colloquium on Structural Information and Communication Complexity*, volume LNCS 4056, pages 29–43, 2006.

6. X. Défago and A. Konagaya. Circle Formation for Oblivious Anonymous Mobile Robots with No Common Sense of Orientation. In *Workshop on Principles of Mobile Computing*, pages 97–104, 2002.

7. E. W. Dijkstra. *Selected Writings on Computing: A Personal Perspective*, pages 34–35. Springer, New York, 1982.

8. P. Flocchini, G. Prencipe, and N. Santoro. Self-Deployment Algorithms for Mobile Sensors on a Ring. Technical Report TR-2006-02, University of Ottawa, 2006.

9. P. Flocchini, G. Prencipe, N. Santoro, and P. Widmayer. Hard Tasks for Weak Robots: The Role of Common Knowledge in Pattern Formation by Autonomous Mobile Robots. In *Proc. 10th International Symposium on Algorithm and Computation*, pages 93–102, 1999.

10. P. Flocchini, G. Prencipe, N. Santoro, and P. Widmayer. Pattern Formation by Autonomous Robots Without Chirality. In *Proc. 8th Int. Colloquium on Structural Information and Communication Complexity*, pages 147–162, June 2001.

11. N. Heo and P. K. Varshney. A Distributed Self Spreading Algorithm For Mobile Wireless Sensor Networks. In *In Proceedings IEEE Wireless Communication and Networking Conference*, volume 3, pages 1597–1602, 2003.

12. A. Howard, M. J. Mataric, and G. S. Sukhatme. An Incremental Self-deployment Algorithm for Mobile Sensor Networks. *Autonomous Robots*, 13(2):113–126, 2002.

13. A. Howard, M. J. Mataric, and G. S. Sukhatme. Mobile Sensor Network Deployment Using Potential Fields: A Distributed, Scalable Solution to The Area Coverage Problem. In *In Proceedings of the 6th International Symposium on Distributed Autonomous Robotics Systems (DARS'02)*, pages 299–308, 2002.

14. B. Katreniak. Biangular Circle Formation by Asynchronous Mobile Robots. In *Proc. of the 12th Int. Colloquium on Structural Information and Communication Complexity*, pages 185–199, 2005.

15. L. Loo, E. Lin, M. Kam, and P. Varshney. Cooperative Multi-Agent Constellation Fonnation Under Sensing and Communication Constraints. *Cooperative Control and Optimization*, pages 143–170, 2002.

16. G. Prencipe. The Effect of Synchronicity on the Behavior of Autonomous Mobile Robots. *Theory of Computing Systems*, 38:539–558, 2005.

17. S. Samia, X. Défago, and T. Katayama. Convergence Of a Uniform Circle Formation Algorithm for Distributed Autonomous Mobile Robots. In *In Journés Scientifiques Francophones (JSF), Tokio, Japan*, 2004.

18. K. Sugihara and I. Suzuki. Distributed Algorithms for Formation of Geometric Patterns with Many Mobile Robots. *Journal of Robotics Systems*, 13:127–139, 1996.

19. I. Suzuki and M. Yamashita. Distributed Anonymous Mobile Robots: Formation of Geometric Patterns. *Siam J. Computing*, 28(4):1347–1363, 1999.

20. O. Tanaka. Forming a Circle by Distributed Anonymous Mobile Robots. Technical report, Department of Electrical Engineering, Hiroshima University, Hiroshima, Japan, 1992.

21. G. Wang, G. Cao, and T. La Porta. Movement-assisted Sensor Deployment. In *In Proceedings of IEEE INFOCOM*, volume 4, pages 2469–2479, 2004.

22. Y. Zou and K. Chakrabarty. Sensor Deployment and Target Localization in Distributed Sensor Networks. *ACM Transactions on Embedded Computing Systems*, 3(1):61–91, 2004.

Minimizing Interference of a Wireless Ad-Hoc Network in a Plane

Magnús M. Halldórsson[1],* and Takeshi Tokuyama[2],**

[1] Dept. of Computer Science, Faculty of Engineering, University of Iceland,
IS-107 Reykjavik, Iceland
mmh@hi.is
[2] Graduate School of Information Sciences, Tohoku University, Sendai, 980-8579
Japan
tokuyama@dais.is.tohoku.ac.jp

Abstract. We consider the problem of topology control of a wireless ad-hoc network on a given set of points in the plane, where we aim to minimize the maximum interference by assigning a suitable transmission radius to each point. By using computational geometric ideas and ϵ-net theory, we attain an $O(\sqrt{\Delta})$ bound for the maximum interference where Δ is the interference of a uniform-radius ad-hoc network. This generalizes a result given in [8] for the special case of highway model (i.c., one-dimensional problem) to the two-dimensional case. We also give a method based on quad-tree decomposition and bucketing that has another provable interference bound in terms of the ratio of the minimum distance to the radius of a uniform-radius ad-hoc network.

1 Introduction

Mobile wireless ad-hoc networks are an important subject in recent studies on communication networks. In a popular model, each mobile device is considered as a point (called *node*) in the Euclidean plane, and each node has a disk of a given transmission radius. Two nodes can communicate with each other if they are located within each other's disks.

The transmission radius is a monotone function of the electric power given to the node, which we assume to be a controllable parameter. Topology control involves assigning a suitable transmission radius to each node to form a connected network while minimizing some non-decreasing objective function of the radii. The most frequently studied objective is to minimize the power consumption, or the sum of the electric power given to the nodes. Making disks small has also another benefit, that is, to reduce the *interference*. Interference at a node is the number of disks containing it, and high interference increases the probability of packet collision of packets. Therefore, it is desirable to keep a low interference at every node.

---

* Partially supported by grants of the Icelandic Research Fund.
** Partially supported by the project *New Horizons in Computing*, Grant-in-Aid for Scientific Research on Priority Areas, MEXT Japan.

S. Nikoletseas and J.D.P. Rolim (Eds.): ALGOSENSORS 2006, LNCS 4240, pp. 71–82, 2006.

Topology control for minimizing interference is bound to be a different task from that of minimizing energy. Traditionally, this has been addressed implicitly by reducing the density of the communication graph. Burkhart et al. [2], however, showed that low interference is not implied by sparseness. Also, that networks constructed from nearest-neighbor connections can fail dismally to bound the interference. On the other hand, they gave experimental results that indicate that graph spanners help reduce interference in practice. Their work prompted the explicit study of interference minimization. Moscibroda and Wattenhofer [6] gave nearly tight approximation algorithms that bound the *average* interference of nodes.

The recent work of Rickenbach et al. [8] is the starting point of our study. They introduced the problem of bounding the maximum interference at a node, and gave algorithms for the special case where all the points are located on a line, called the *highway model*. Their algorithm constructs a network with an $O(\sqrt{\Delta})$ interference, where Δ is the interference of a uniform radius network, while it is shown that there exists an instance that requires $\Omega(\sqrt{n})$ interference. They also showed that the better one of a naive network and the above $O(\sqrt{\Delta})$ interference network attains a $O(\Delta^{1/4})$ approximation ratio.

For the two-dimensional problem, analogous results have not been reported yet (to the authors' knowledge). In this paper, we show that we can construct a network with an $O(\sqrt{\Delta})$ interference for any point set in the plane, extending the theory of [8] to the planar case (and even for any constant-dimensional space). We also give a network with an $O(\log(R/d))$ interference, where d is the minimum distance between points and R is the minimum radius of a uniform-radius network to attain connectivity. Our results rely on computational geometric tools such as local neighbor graphs, ϵ-nets, and quad-tree decompositions.

2   Mathematical Formulation and Terminology

We are given a set $V = \{\mathbf{v}_1, \mathbf{v}_2, \ldots, \mathbf{v}_n\}$ of points in a plane. For each $\mathbf{v}_i$, we assign a positive real number $r(\mathbf{v}_i)$ called the *transmission radius*. This can be considered as a *radius assignment* function

$$r : V \to \mathbb{R}_{>0}.$$

Consider the set $\mathcal{D} = \{D_1, D_2, \ldots, D_n\}$ of disks, where D_i has radius $r(\mathbf{v}_i)$ and its center at $\mathbf{v}_i$.

We define a wireless network on V, that is the graph $G(\mathcal{D}) = (V, E)$, where we have an undirected edge $(\mathbf{v}_i, \mathbf{v}_j)$ if and only if $\mathbf{v}_i \in D_j$ and $\mathbf{v}_j \in D_i$. In other words, $\mathbf{v}_i$ and $\mathbf{v}_j$ can directly communicate since they are within the transmission radius of each other. We say that the wireless network $G(\mathcal{D})$ is *feasible* iff it is connected.

The interference of $\mathcal{D}$ at a point $\mathbf{p}$ is the number of disks in $\mathcal{D}$ covering $\mathbf{p}$. That is,

$$I(\mathcal{D}, \mathbf{p}) = |\{i : \mathbf{p} \in D_i\}|.$$

The *interference* of a wireless network $G(\mathcal{D})$ is [1]

$$\max\{I(\mathcal{D}, \mathbf{p}) | \mathbf{p} \in \mathbb{R}^2\}.$$

The *interference minimization problem* is to find a radius assignment r to give a feasible network with the minimum interference.

One natural approach is to increase all radii uniformly until the the graph becomes connected. Let R_{min} be the infimum of the radius such that the network becomes connected, and refer to the network with all radii set to R_{min} as the *uniform-radius* network. Let Δ denote the interference of the uniform-radius network.

Although the problem is clearly an NP-optimization problem, it seems to be very difficult to find the optimal wireless network. Indeed, even the special case where all points V are located on a line (highway model) is considered to be difficult (although NP-hardness result is not known). Thus, we seek for a practical solution with some theoretical quality guarantee, in particular an upper bound of the interference or an approximation ratio to the optimal solution.

2.1   Review for the Highway Model

We briefly review some results for the highway model given by Rickenbach *et al.* [8]. Suppose that points of V are located on the x-axis in the sorted order with respect to their x-values.

Then, a naive method is to set $r(i) = \max(d(\mathbf{v}_i, \mathbf{v}_{i-1}), d(\mathbf{v}_i, \mathbf{v}_{i+1}))$ for $i = 1, 2, \ldots, n$, where we set $\mathbf{v}_0 = \mathbf{v}_1$ and $\mathbf{v}_{n+1} = \mathbf{v}_n$. It is easy to observe that $G(\mathcal{D})$ associated with this radius function is feasible: the network is called the *linear network*. Unfortunately, there is an example named *exponential chain* instance for which the linear network poorly performs. In the exponential chain, the points satisfies that $d(\mathbf{v}_i, \mathbf{v}_{i+1}) = 2^i$ for $i = 1, 2, \ldots n - 1$, and it is easy that the interference of the point v_1 is $n - 1$ in the linear network.

We can use a *hub-connected* network to reduce the worst-case interference. The general idea is as follows: We find a subset $W \subset V$ of points called *hubs* and first construct the linear network of hubs. Then, for each $\mathbf{v} \in V \setminus W$, we set

$$r(\mathbf{v}) = \min_{\mathbf{w} \in W} d(\mathbf{v}, \mathbf{w});$$

namely, $\mathbf{v}$ connects to its nearest hub. If we select every $\sqrt{n}$-th points in V as a hub, we have a set W of cardinality $\sqrt{n}$, and it is shown that $I(G(\mathcal{D})) = O(\sqrt{n})$ for this network. It has been shown that the interference is $\Omega(\sqrt{n})$ for the exponential chain, thus the hub-connected network is worst-case optimal. However, for each given instance, we can often design a network with a better interference. Indeed, there is a construction that $I(G(\mathcal{D})) = \sqrt{\Delta}$.

---

[1] We can also consider the version where we only consider interference at points of V, not all points in the plane. The results of this paper carry immediately over to that model.

2.2   Two-Dimensional Analogue of the Linear Network

Although the linear network performs poorly in the worst case, it is a basic structure that can also be constructed in a distributed fashion. That is, each point can connect to its right and left neighbors without the need of global information.

The first task is to extend this notion to the two-dimensional case, where we do not have clear definitions of the left and right neighbors. If we sort the points with respect to x-coordinate, and each point connects to the nearest neighbor with respect to the x-coordinate, we can obtain a feasible network. However, this ignores the y-coordinate, and usually gives a very bad network. Instead, we would like to use the Euclidean distance to measure the proximity of points.

Indeed, a network in which each node establishes (two-way) connection with its nearest neighbor is called a *nearest-neighbor forest*. The nearest-neighbor forest need not be connected, however, and we want give a connected network based on it. The minimum spanning tree MST(S) might be a direct two-dimensional analogue of the linear network. The wireless version is WMST(S) in which each node $\mathbf{p}_i$ has the radius $\max_{\mathbf{q}:(\mathbf{p}_i,q)\in \text{MST}(S)} d(\mathbf{p}_i,q)$. Constructing a minimum spanning tree explicitly requires some global information; hence, we prefer graphs of a more local nature.

We briefly explain the *local neighborhood graph* (LNG) [9], since it inspires the construction of our hub-structure network given later.

For each point $\mathbf{p} \in \mathbb{R}^2$, we divide the plane into six cones $R_1(\mathbf{p}), R_2(\mathbf{p}), \ldots,$ $R_6(\mathbf{p})$, where $R_k(\mathbf{p})$ is the region such that the argument angle about $\mathbf{p}$ is in the range $[\frac{(k-1)\pi}{3}, \frac{k\pi}{3})$.

Let $nb_k(\mathbf{p}, V)$ be the nearest point to $\mathbf{p}$ in $V \cap R_k(\mathbf{p})$. See Figure 1. The local neighbor graph LNG(V) is the graph connecting each $\mathbf{v} \in V$ to its six local neighbors.

The following elementary fact is important and will be used to show an interference bound for our network given later.

Lemma 1. *Suppose that $\mathbf{u}$ and $\mathbf{v}$ are in $R_k(\mathbf{p})$ and $d(\mathbf{p}, \mathbf{u}) \leq d(\mathbf{p}, \mathbf{v})$. Then, $d(\mathbf{u}, \mathbf{v}) < d(\mathbf{p}, \mathbf{v})$.*

Proof. Straightforward from the fact that the diameter (distance between farthest pair of points) of a fan with the angle $\pi/3$ equals the radius of the circle.

The above lemma leads to the following fact [9], although we do not give a proof since we do not use this fact explicitly in the rest of the paper.

Lemma 2. LNG(V) *contains* MST(V). *Consequently, it is connected.*

Let $N_1(\mathbf{v}_i) = \{nb_k(\mathbf{v}_i, V)|1 \leq k \leq 6\}$ and $N_2(\mathbf{v}_i) = \{\mathbf{w} \in V | \mathbf{v}_i \in N_1(\mathbf{w})\}$. If we set $r_i = \max\{d(\mathbf{v}_i, \mathbf{q}) | \mathbf{q} \in N_1(\mathbf{v}_i) \cup N_2(\mathbf{v}_i)\}$ for each $i = 1, 2, \ldots n$, we have a wireless network WLNG(V) that has LNG(V) as a subgraph.

We remark that WLNG(V) can be constructed locally: Each node increases its radius (up to a given limit) and sends a message until it receives acknowledgement from the local neighbor in each of six cones, and sends a connection

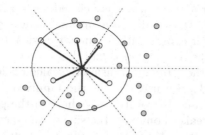

Fig. 1. Local neighbors of a point and a disk connecting them

request to each local neighbor. Then, each node that receives connection request increases the radius such that it can reach the sender. We remark that this method has the defect that we need to set the limit radius, since if there is an empty cone, we have to detect and ignore it to avoid increasing the radius to infinity.

2.3 Hub-Connected Network with $O(n^{1/2})$ Interference

It is known that we can make a bad instance for which any network containing the nearest-neighbor forest has an $\Omega(n)$ interference while there exists a network with a constant interference for the instance [8]. Thus, if every node connects to its nearest neighbor, we can obtain neither a (nontrivial) absolute interference bound nor a rational bound to the optimal interference for each instance.

In order to attain a better interference bound, we consider a hub-connected network, where we select a subset W of V as the set of hubs. We construct WMST(W) as the core of the network, and propagate the connection around the core such that every vertex $\mathbf{v} \in V \setminus W$ is connected to the nearest hub to it. Note that we may use any connected network on W (e.g., $WLNG(W)$) as the core instead of WMST(W) in order to attain our main theoretical result: what is important is the choice of W.

Hub Selection Using an ϵ-net. We apply ϵ-net theory to define the set W of hubs. Consider a family $\mathcal{R}$ of regions in the plane. Given a set V of n points, the pair $(V, \mathcal{R})$ is called a *range space*. An ϵ-net of the range space $(V, \mathcal{R})$ is a subset $S \subset V$ such that any region $R \in \mathcal{R}$ that contains at least ϵn points of V must contain at least one point of S. Intuitively, an ϵ-net is a uniformly distributed sample of V where the uniformity is measured by using the family $\mathcal{R}$ of regions.

The following theory (although readers need not be familiar with it) has many applications such as computational geometry [1] and learning theory: The *Vapnik-Chervonenkis*-dimension (VC dimension) of a range space is the largest size of a subset $A \in V$ such that all subsets of A are attained as an intersection of A and a region in $\mathcal{R}$. If VC dimension is low (say, a constant), we can always have a small ϵ-net (see [4] for example).

Here, we consider a range space associated with a family of sectors of disks. Consider a unit disk D, and divide it into six cone sectors P_k $(k = 1, 2, \ldots, 6)$ by the three diameter chords with argument angles 0, $\pi/3$, and $2\pi/3$. That is, $P_k = \{x \in D | (k-1)\pi/3 \leq arg(x) < k\pi/3\}$.

The family $\mathcal{P}_k$ is the set of all translated/scaled copies of P_k. We consider the family $\mathcal{P} = \cup_{1 \leq k \leq 6} \mathcal{P}_k$. Intuitively, it is the family of "1/6 piece of pies" in six rotated positions of any size located anywhere in the plane.

First, we give a weaker bound for the size of an ϵ-net of $\mathcal{P}$. Although this will be slightly improved later, the following result is useful since we do not need any complicated algorithm to find the ϵ-net.

Theorem 1. *A random sample of size $c\epsilon^{-1} \log \epsilon^{-1}$ becomes an ϵ-net for $\mathcal{P}$ with high probability if c is a sufficient large constant.*

Proof. If we construct an ϵ-net for each of $\mathcal{P}_k$, their union becomes an ϵ-net of $\mathcal{P}$. Thus, it suffices to show the existence of an ϵ-net of size $O(\epsilon^{-1} \log \epsilon^{-1})$ for each of $\mathcal{P}_k$. We have the theorem from the general theory of ϵ-nets [3,4] of range spaces.

A family $\mathcal{R}$ of regions is said to be a family of *pseudo-disks* if for any noncollinear three points in the plane, there exists a unique $R \in \mathcal{R}$ such that those three points are on the boundary of R. The following better bound is known for a family of pseudo-disks.

Theorem 2. *[5] For any point set V, there is an ϵ-net of size $O(1/\epsilon)$ for a family of pseudo-disks.*

Consider the family $\mathcal{P}_k$ for $k = 1, 2, \ldots, 6$, say, $k = 1$. It is easy to see that for any noncollinear three points in the plane, there exists at most one $P \in \mathcal{P}_1$ such that the triple of points are on the boundary of P. Thus, $\mathcal{P}_1$ has a property that is very similar to pseudo-disks, but there may be triplets of points such that there is no $P \in \mathcal{P}_1$ such that the boundary of P goes through them. Nevertheless, we have the following theorem:

Theorem 3. *There exists an ϵ-net of V of size $O(1/\epsilon)$ for $\mathcal{P}$, and we can compute one in polynomial time.*

This theorem is of independent interest in the area of computational geometry. Since it probably requires too much geometric knowledge for a non-specialist to follow, we give (an outline of) the actual construction of such an ϵ-net later in a separate section.

The Hub-Connected Network. The construction is as follows: We first compute an $\sqrt{n^{-1}}$-net W of V such that the size of W is $O(\sqrt{n})$, which can be obtained by using Theorem 3 by setting $\epsilon = \sqrt{n^{-1}}$. Then, we form the wireless network WMST(W) (indeed, any connected network is fine for our purpose). Let $r_0(w)$ be the transmission radius of $w \in W$ in WMST(W).

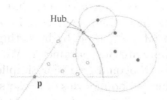

Fig. 2. No disk around a point outside the fan can reach **p**

We call the elements of W *hubs*. Then, for each $v \in V \setminus W$, we find its nearest hub denoted by $hub(v)$. We set $r(v) = d(v, hub(v))$. For each hub $w \in W$, define the set $N(w) = \{v \in V \setminus W | hub(v) = w\}$. We set $r(w) = \max\{r_0(w), \max_{v \in N(w)} d(v, w)\}$ for each $w \in W$. We have determined r for each elements of v, and thus we obtain a wireless network GHUB(V).

Lemma 3. GHUB(V) *is connected.*

Proof. Since WMST(W) is connected, the induced subgraph of GHUB(V) by W is connected. Since other nodes are all connected to nodes in W, GHUB(V) is connected.

Theorem 4. *The interference of* GHUB(V) *is* $O(\sqrt{n})$.

Proof. Let c be a suitable constant such that $|W| < c\sqrt{n}$. We claim that any point $\mathbf{p} \in \mathbb{R}^2$ is covered by at most $(c + 6)\sqrt{n}$ disks, or, more precisely, by $6\sqrt{n}$ disks except those around elements of W. Consider the cusp $R_1(\mathbf{p})$ whose argument angle interval is $[0, \pi/3]$. Because of symmetry, it suffices to show that at most $\sqrt{n}$ points in $R_1(\mathbf{p})$ can contain $\mathbf{p}$ in their disks. If there is no hub in $R_1(\mathbf{p})$, then $R_1(\mathbf{p})$ cannot contain more than $\sqrt{n}$ points because W is a $\sqrt{n}^{-1}$-net, and we are done. Otherwise, we can assume there is at least one hub in $R_1(\mathbf{p})$ (see Figure 2). Let $\mathbf{w}$ be the nearest hub to $\mathbf{p}$ in $R_1(\mathbf{p})$. We draw a circle C of radius $d(\mathbf{p}, \mathbf{w})$ around $\mathbf{p}$, and let P be the region (i.e., piece of pie) obtained as the intersection of the interior of C and $R_1(\mathbf{p})$. Since P does not contain a hub in its interior, P can contain at most $\sqrt{n}$ elements of V. Consider any point $\mathbf{x} \in V$ in $R_1(\mathbf{p}) \setminus P$. Then, it follows that $d(\mathbf{x}, \mathbf{w}) < d(\mathbf{x}, \mathbf{p})$ from Lemma 1 (here, it is crucial that the angle of a fan is $\pi/3$). Since $r(\mathbf{x})$ is the distance to its nearest hub, $r(\mathbf{x}) \leq d(\mathbf{x}, \mathbf{w}) < d(\mathbf{x}, \mathbf{p})$. Thus, $\mathbf{p}$ is not in the disk of $\mathbf{x}$. This completes the proof.

Note that if we use the weaker ϵ-net obtained by random sampling, we set $\epsilon = \sqrt{n^{-1} \log n}$ to have a network with an interference $O(\sqrt{n \log n})$.

3   A Network with $O(\sqrt{\Delta})$ Interference

Let us consider the uniform-radius network G_0 in which each disk has the same radius R_{min}. Recall that Δ is the interference of G_0. Although Δ can become as large as $\Omega(n)$, it can in practice be much smaller than n, or even $\sqrt{n}$. We show a construction of a network where the interference is $O(\sqrt{\Delta})$.

We use a standard localization method by bucketing. By scaling, we can assume that $R_{min} = 1$ to eliminate one parameter. We partition the plane into unit square buckets by an orthogonal grid. For simplicity of argument, we assume that there are no points on boundaries of buckets; this assumption is easy to remove. G_0 can connect a point $v \in B$ to points in bucket B or its eight neighbors. We say that two buckets B and B' are adjacent if there exists $v \in B$ and $v' \in B'$ such that the edge (v, v') is in G_0.

Lemma 4. *1. For each B, an adjacent bucket must be one of its eight neighbors in the grid.*
  2. Each bucket contains $O(\Delta)$ points.

Proof. (i) is obvious, since the distance from any point in B to any bucket other than the eight neighbors is more than 1. For (ii), suppose that a bucket contains more than 4Δ points. We refine the buckets into four sub-buckets of size 0.5×0.5. One of the sub-bucket contains more than Δ points, and the center of the sub-bucket is covered by the unit disk about each point in its sub-bucket. This contradicts that the interference of G_0 is Δ.

Our construction is as follows: First, in each bucket B, we give a network with interference $O(\sqrt{\Delta})$ by using the construction given in the previous subsection, and set the radius of each point accordingly. Note that none of the disks in the construction has a radius larger than $\sqrt{2}$. Second, for each adjacent pair B, B' of buckets, select exactly one edge $(v, v') \in G_0$ connecting them. We call v and v' *connectors*. We enlarge the radius of each connector to 1 (if its current radius is less than 1) .

Now, we have defined all the radii, and accordingly we have a network $\mathrm{LHUB}(V)$.

Theorem 5. *The network $\mathrm{LHUB}(V)$ is connected, and its interference is $O(\sqrt{\Delta})$.*

Proof. The network is connected within each bucket, and the connection between buckets is same as G_0. Thus, it is connected. For each point p, it is interfered by points of at most 21 buckets (the neighbor buckets of Manhattan distance at most two), since the radius of the largest disk is $\sqrt{2}$. Each bucket contribute only $O(\sqrt{\Delta})$, excluding connectors. Also, there are only constant number of connectors in these buckets. Thus, we have the theorem.

4   A Hierarchical Construction

The GHUB network has two layers: hubs and others. LHUB network has three layers: connectors, hubs in buckets, and others. One may feel that we may have a better structure if we increase the number of layers. If we measure the worst-case interference by using the input size n or Δ, it is not possible to improve the worst case interference, since there is a lower bound of $\Omega(\sqrt{\Delta})$ even in the highway (i.e., one-dimensional) model. However, this can be advantageous in practice as we see if we measure the interference using a different parameter.

Let d be the minimum distance between two points in V. Below, we will give a network whose interference ratio is $O(\log(R_{min}/d))$, where R_{min} is the radius to give the uniform-radius network. As before, we scale the problem such that $R_{min} = 1$.

The same localization method works, and we can assume that all points are located in a unit square. Our approach is based on quad-tree decomposition. We adopt the convention that each square in the quad-tree decomposition includes its lower edge and its right edge, together with its lower two corner vertices.

We continue the following process from $k = 0$, where $U(S) = V$ if $k = 0$:

Quad-tree decomposition process: Given a square S of size $2^{-k} \times 2^{-k}$ and a set $U(S) \subset V \cap S$ do the following.

1. If $U(S) = \emptyset$, terminate the process.
2. Otherwise, select a representative point $\mathbf{p}(S) \in V(S)$ arbitrarily, and remove $\mathbf{p}(S)$ from $U(S)$.
3. Partition S into four quadrants of size $2^{-(k+1)} \times 2^{-(k+1)}$. The point set $U(S)$ is partitioned accordingly. The at most four non-empty quadrants obtained are called *children* of S.
4. Apply the process iteratively to each child.

We call S' the parent of S if S is one of the children of S', and denote $S' = parent(S)$. We also say that $\mathbf{p}(S)$ is a child (resp. parent) of $\mathbf{p}(S')$ if S is a child (resp. parent) of S'. For the representative point $\mathbf{p}(S)$ of S, we set $r(\mathbf{p}(S)) = \max\{diag(S), d(\mathbf{p}(S), \mathbf{p}(parent(S)))\}$, where $diag(S)$ is the length of the diagonal of the square S. Thus, we have assigned a radius to each point of V, and have a network QUAD(V).

Theorem 6. QUAD(V) *is connected, and its interference is* $O(\log d^{-1})$, *where d is the minimum distance between points of V.*

Proof. Since $r(\mathbf{p}(S)) \geq diag(S)$, the disk of $\mathbf{p}(S)$ contains all its children. Also, $r(\mathbf{p}(S)) \geq d(\mathbf{p}(S), \mathbf{p}(parent(S)))$ means that the disk also contains its parent. Thus, the points are connected via the tree structure of the parent-child relation.

Now, let us analyze the interference at a point $\mathbf{p}$. There are at most $O(\log d^{-1})$ different sizes of squares in the quad tree decompositions, since the diagonal length of the parent square of a smallest square must be at least d (otherwise, it can contain only one point). Consider a bucket size 2^{-k}, and analyze how many representative points of such buckets can interfere with $\mathbf{p}$. The radius $r(\mathbf{p}(S))$ of a representative point of a square S of this size is at most $2^{-k+1}\sqrt{2}$, since the distance from the representative point to any point in the parent square is at most $diag(parent(S)) = 2^{-k+1}\sqrt{2}$. Thus, S can interfere with $\mathbf{p}$ only if S intersects with the circle of radius $2^{-k+1}\sqrt{2}$ about $\mathbf{p}$. It is easy to see that there are only a constant number of such squares of this size. Thus, the interference at $\mathbf{p}$ is $O(\log d^{-1})$.

In a practical implementation, we should apply a routine to shrink each disk as much as possible while keeping the connection to its parent and children.

5　Construction of a Small-Size ϵ-net

Here, we give an outline of a proof of Theorem 3. It suffices to show the following:

Theorem 7. *There exists an ϵ-net of V of size $O(1/\epsilon)$ for $\mathcal{P}_1$.*

We follow the argument of [5] with a (minor) modification. For simplicity, we assume that no two points of V lie on a horizontal line, a vertical line, or a line with the argument angle $\pi/3$. We call a member of $\mathcal{P}_1$ a *fan* in this section. For a fan P, we define $Int(P)$ and $cl(P)$ to be its interior and closure. Let $\partial(P) = cl(P) \setminus Int(P)$ be the boundary of its closure.

Given a subset $S \subset V$, a pair $(\mathbf{p}, \mathbf{p}')$ of points in S is extremal in S if, for any number $N > 0$, there is a fan P such that the area of P is larger than N, $Int(P) \cap S = \emptyset$ and $\{\mathbf{p}, \mathbf{p}'\} \in \partial(P)$.

We add a set X of three "extra" points $\mathbf{q}_1, \mathbf{q}_2, \mathbf{q}_3$ to V. Let ℓ_1 be a horizontal line that contains V in its lower halfplane. Let ℓ_2 be a line of argument angle $\pi/3$ that contains V in its upper halfplane. The points $\mathbf{q}_1$ and $\mathbf{q}_2$ are on the line ℓ_1, and the x-coordinate value of $\mathbf{q}_1$ (resp. $\mathbf{q}_2$) is sufficiently small (resp. large). The point $\mathbf{q}_3$ is on the line ℓ_2 and its y-coordinate value is sufficiently small. We can take these three points sufficiently far from V such that X satisfies the following conditions:

1. The triangle spanned by X contains all points of V.
2. For any fan P, we have another fan $P' \subseteq P$ such that $P' \cap V = P \cap V$, and $P' \cap X = \emptyset$.
3. For any pair of points in X, there is a fan P containing them on the boundary and containing no other points of V in it.
4. For any extremal pair $(\mathbf{p}, \mathbf{p}')$ of a subset S of V, we have a fan P with the largest size such that $Int(P) \cap (S \cup X) = \emptyset$ and $\{\mathbf{p}, \mathbf{p}'\} \in \partial P$. Note that one or more points of X lie on the boundary of P, and intuitively, X prevents $(\mathbf{p}, \mathbf{p}')$ to be an extremal pair in $S \cup X$.

Now, we fix $S \in V$ and consider $\tilde{S} = S \cup X$. We say a fan P an *empty fan* if it contains no point of $\tilde{S}$ in its interior. A pair of points $(\mathbf{p}, \mathbf{p}')$ is called a Voronoi pair if there exists an empty fan P containing $\mathbf{p}$ and $\mathbf{p}'$ in its boundary. Let $DT(S)$ be the graph whose node set is $\tilde{S}$ and edge set E is the set of all Voronoi pairs.

Lemma 5. *$DT(S)$ is connected, and gives a triangulation with the vertex set $\tilde{S}$ in the triangle spanned by X.*

Proof. It is easy to show that no pair of edges intersect each other using the fact that two fans intersect each other such that boundary curves intersect at most twice. Given an emtpy fan P containing a Voronoi pair $(\mathbf{p}, \mathbf{p}')$ on its boundary, we can grow P keeping the Voronoi pair on the boundary until we have another point $\mathbf{p}''$

in $\tilde{S}$ on its boundary. Then, we have a triangle $\mathbf{p}, \mathbf{p}', \mathbf{p}''$ in $DT(S)$ consisting of three edges. It can be shown (by case study) that if the Voronoi pair does not contain a point in X, we have exactly one such triangle in each side of the edge. Thus, we can show both the connectivity and triangulation property.

$DT(S)$ is called the generalized Delauney triangulation of S. For each triangle in $DT(S)$, the unique fan P containing three vertices of triangles on its boundary is called the *Voronoi fan* to the triangle. Note that a Voronoi fan contains no point of S in its interior.

The construction of [5] is as follows: Let $\delta = \epsilon/4$. We greedily find a maximal family of disjoint subsets $\{S_1, S_2, \ldots, S_k\}$ of V such that $|S_i| = \delta n$ and there exists a fan P_i such that $P_i \cap V = S_i$.

Let $S = \cup_{i=1}^k S_i$, and we make $DT(S)$. By definition, any fan P containing δn or more points of V must contain a point of S. Thus, for each triangle in $DT(S)$, there are at most δn points of V in the Voronoi fan. Moreover, the subgraph of $DT(S)$ induced by S_i is connected, and each Voronoi fan corresponding to a triangle in the induced subgraph contains no point of V in its interior.

We use $k + 3$ colors to give a mutually different color to each set S_i and also each of three point of X. The points in $V \setminus S$ are colorless. We give corresponding colors to vertices of $DT(S)$. For two colors (c_1, c_2), a triangle is called (c_1, c_2)-colored if its vertices use exactly the two colors.

For a fixed pair (c_1, c_2) of colors, the adjacency graph of the set of all (c_1, c_2)-colored triangles has neither a branching node (i.e., a node with degree three or more), nor a cycle: This can be shown by using the fact that S_i is intersection of a fan (a convex region) and V. Thus, the set of (c_1, c_2)-colored triangles is divided into maximal connected chains of triangles called *corridors*.

Lemma 6 ([5]). *There are $O(k)$ corridors.*

The corridors are refined into sub-corridors such that each sub-corridors has at most δn points of V (indeed, they are colorless) in its triangles. The vertex set of subcorridors C consists of two monochromatic chains (possibly degenerated to points) in $D(S)$, and thus they have at most four endpoints.

Let Z be the set of all endpoints of all subcorridors in $DT(S)$.

Theorem 8. *Z is an ϵ-net of $V \cup X$, and its size is $O(1/\epsilon)$.*

Proof. The number of sub-corridors is $O(k + \frac{1}{\delta}) = O(1/\epsilon)$. Thus $|Z| = O(1/\epsilon)$. Consider any fan P containing more than ϵn points of $V \cup X$. We assume that P contains no point of Z and derive contradiction. P can contain no monochromatic chain in its iterior since it does not have a point in Z. If the fan P cuts both monochromatic chains of a subcorridor, $P \cup V$ must be contained in two colored sets (by an argument given in [5], which we omit in this paper), and contradict to the assumption, since each monochromatic set contains $\epsilon n/4$ points. Thus, for each subcorridor, P can only cut one of its mornochromatic chain. This implies that $P \cup V$ is monochromatic (plus some colorless points), also, it can be shown that the number of colorless points in it is at most $3\epsilon n/4$ (we omit details). Thus, we have contradiction.

We finally show that $Z \setminus X$ is an ϵ'-net of V if $\epsilon < \epsilon' < 2\epsilon$. Indeed, suppose we have a fan P that contains $\epsilon'n$ points of V but no point in $Z \setminus X$. Thus, it must contain one or more points of X. We can shrink P such that only the points of X go outside of it. This new fan contains $\epsilon'n - 3$ points of V and contains no point in Z. Thus, this contradicts the fact that Z is an ϵ-net of $V \cup X$.

6   Concluding Remarks

The theory can easily be generalized to any constant dimensional space, except that we only know a $O(\epsilon^{-1} \log^{-1} \epsilon^{-1})$ bound for ϵ-nets of the higher dimensional analogues of "the range space of pies".

Practically, we can improve the method in many ways. For example, in the construction of QUAD(V), we can stop the partitioning if $|U(S)| = 1$, and else partition $U(S)$ without selecting a representative point until there are at least two empty buckets. Also, we can mix the two methods: In each square S, we can replace the structure of QUAD(S) network within S by LHUB(S), if it gives a better interference.

There are several open problems: It is observed that an $\Omega(\sqrt{\log(R_{min}/d)})$ lower bound is attained by the "exponential chain instance" in the highway model. We conjecture that this lower bound is tight, although we currently only have an $O(\log(R_{min}/d))$ upper bound given in this paper. Moreover. for the highway model, the better one of linear network and hub network attains $O(\Delta^{1/4})$ approximation ratio to the optimal network. For the two-dimensional case, an analogous result has not been obtained yet.

References

1. P. K. Agarwal, Range Searching, Chapter 31 of *Handbook of Discrete and Computational Geometry* (ed. J. E. Goodman and J. O'Rourke), CRC Press 1997, pp. 575-598.
2. M. Burkhart, P. von Rickenbach, R. Wattenhofer, A. Zollinger: Does Topology Control Reduce Interference? Proc. MobiHoc 2004, 9-19 (2004).
3. D. Haussler and E. Welzl, ϵ-Nets and Simplex Range Queries, *Discrete & Compu. Geom.* **2** (1987), 237-256.
4. J. Matoušek, *Geometric Discrepancy, An Illustrated Guide*, Springer Verlag, 1999.
5. J. Matoušek, R. Seidel, E. Welzl, How to Net a Lot with Little: Small ϵ-Nets for Disks and Halfspaces, *Proc. 6th Symp. on Computational Geometry,*(1990), 16-22.
6. T. Moscibroda, R. Wattenhofer, Minimizing Interference in Ad Hoc and Sensor Networks, *Proc. DIALM-POMC Joint Workshop* 2005, 24-33.
7. F. P. Preparata and M. I. Shamos, *Computational Geometry, an Introduction*, Texts and Monographs in Computer Science, Springer Verlag, 1988 (2nd ed).
8. P. von Rickenbach, S. Schmid, R. Wattenhofer, A. Zollinger: A Robust Interference Model for Wireless Ad-Hoc Networks, Proc. IPDPS 2005 (2005).
9. A. C.-C. Yao, On Constructing Minimum Spanning Trees in k-Dimensional Spaces and Related Problems, *SIAM J. Comput.,* **11** (1982), 721-736.

Self-stabilizing Weight-Based Clustering Algorithm for Ad Hoc Sensor Networks

Colette Johnen and Le Huy Nguyen

LRI–Université Paris Sud, CNRS UMR 8623
Bâtiment 490, F91405, Orsay Cedex, France
colette@lri.fr,lehuy@lri.fr

Abstract. Ad hoc sensor networks consist of large number of wireless sensors that communicate with each other in the absence of a fixed infrastructure. Fast self-reconfiguration and power efficiency are very important property on any sensor network management. The clustering problem consists in partitioning network nodes into groups called clusters, thus giving at the network a hierarchical organization. Clustering increases the scalability and the energy efficiency of communication among the sensors. A self-stabilizing algorithm, regardless of the initial system state, converges to a set of states that satisfy the problem specification without external intervention. Due to this property, self-stabilizing algorithms are adapted highly dynamic networks. In this paper we present a Self-stabilizing Clustering Algorithm for Ad hoc sensor network. Our algorithm adapts faster than other algorithms to topology changes.

Keywords: Self-stabilization, Distributed algorithm, Clustering, sensor network.

1 Introduction

An *ad hoc sensor* network consists of a number of sensors spread across a geographical area. Each sensor has wireless communication capability and some level of intelligence for signal processing and networking. Given the large number of nodes and their potential placement in hostile locations, it is essential that the network be able to self-organize; manual configuration is not feasible. Moreover, nodes may fail at any time (either from lack of energy or from physical destruction), and new nodes may join the network. Therefore, the network must be able to reconfigure itself so that it can continue to function. The lifetime of a sensor is determined by the battery life, thereby requiring the minimization of energy expenditure for network management procedure. Therefore, fast self-reconfiguration has to be the main feature of a sensor network management.

Clustering means partitioning network nodes into groups called clusters, giving to the network a hierarchical organization. A cluster is a connected graph composed of a clusterhead and (possibly) some ordinary nodes. Each node belongs to only one cluster. In addition, a cluster is required to obey to certain constraints that are used for network management, routing methods, resource

S. Nikoletseas and J.D.P. Rolim (Eds.): ALGOSENSORS 2006, LNCS 4240, pp. 83–94, 2006.

allocation, etc. The idea of one-level clustering can be easily generalized to a multilevel hierarchical network decomposition. Each node of the network is a level-0 cluster. The level-i clusters are grouped together into level-i+1 clusters, for $i \geq 0$. By dividing the network into non-overlapped clusters, intra-cluster routing is administered by the clusterhead and inter-cluster routing can be done in reactive manner by clusterhead leaders and gateway. Clustering has the following advantages. First, clustering facilitates the reuse of resource, which can improve the system capacity. Clustering-based routing reduces the amount of routing information propagated in the network. Clustering reduces the amount of information that is used to store the network state. The clusterhead will collect the state of nodes in its cluster and built an overview of its cluster state. Distant nodes outside of the cluster usually do not need to know the details of specific events occurring inside the cluster. Hence, an overview of the cluster's state is sufficient for those distant nodes to make control decisions. Clustering is vital for efficient resource utilization and load balancing in large scale networks as sensor networks.

For these reasons, it is not surprising that several distributed clustering algorithms have been proposed during the last few years [1, 2, 3, 4, 5, 6, 7]. The clustering algorithms appeared in [1, 5] build a spanning tree. Then on top of the spanning tree, the clusters are constructed. In these papers, the clusterheads set is not a dominating set (i.e., a processor can be at distance greater than 1 of its clusterhead). Two network architectures for MANET (Mobile Ad hoc Wireless Network) are proposed in [6, 7] where nodes are organized into clusters. The built clusterheads set is an independent (i.e., clusterheads are not neighbors) and also a dominating set. The clusterheads are selected according to the value of their IDs. In [3], a Distributed and Mobility-Adaptive Clustering algorithm, called DMAC, is presented; the clusterheads are selected according to a node's parameter (called *weight*). The higher is the weight of a node, the more suitable this node is for the role of clusterhead. An extended version of this algorithm, called Generalized DMAC (GDMAC), was proposed in [2]. In the latter algorithm, the clusterheads set does not have to be an independent set. This implies that, when, due to mobility of the nodes, two or more clusterheads become neighbors, none has to resign. Thus, the clustering management with GDMAC requires less overhead than the clustering management with DMAC in highly mobile environment. The DMAC and GDMAC algorithms are analyzed respectively in following papers [8, 9], with respect to their convergence time and message complexity. In [4], a weight-based distributed clustering algorithm is presented; also the computation of the node's weight according several parameters (node's degree, transmission power, battery power, ...). In [10, 11] probabilistic clustering constructions for ad hoc sensor network are presented.

A self-stabilizing, when "regardless of its initial state, is guaranteed to arrive at a legitimate state in a finite number of steps". A system which is not self-stabilizing may stay in an illegitimate state forever. The design of self-stabilizing distributed algorithms has emerged as an important research area in recent years [12, 13]. The correctness of self-stabilizing algorithms does not depend on initialization of variables, and a self-stabilizing algorithm converges to some predefined stable state starting from an arbitrary initial one. Self-stabilizing algorithms are thus inherently tolerant to transient faults in the system. Many self-stabilizing algorithms can also adapt dynamically to changes in the network topology or system parameters (e.g., communication speed, number of nodes). A state following a topology changes is seen as an inconsistent state from which the system will converge to a state consistent with the new topology. [14] presents a self-stabilizing algorithm that builds a maximal independent set (i.e., members of the set are not neighbors, and the set cannot contains any other processors). Notice that a maximal independent set is a good candidate for the clusterheads set because a maximal independent set is also a dominating set (i.e., any processor is member of the dominating set or has a neighbor that is member of the set). In [15], a self-stabilizing algorithm that creates a minimal dominating set (i.e., if a member of the set quits the set, the set is not more a dominating set) is presented. Notice that a minimal dominating set is not always an independent set.

Both algorithms DMAC and GDMAC are not self-stabilizing, i.e., they work assuming correct initialization. They cannot cope with the wake up problem. Sensors to conserve energy sleep a large portion of the time. During the sleeping period of a sensor, the network topology may have drastically changed. The sensor has to automatically adapt to the new situation. We present in this paper a self-stabilizing version of DMAC and GDMAC algorithm: they cope with any initial configuration. They also adapt to arbitrary topology changes due to node crash failures, communication link crash failures, node recovering or link recovering, merging of several networks, and so on.

In [16], a self-stabilizing link-cluster algorithm under an asynchronous message-passing system model is presented (no convergence proofs are presented). The definition of cluster is not exactly the same as ours: an ordinary node can be at distance two of its clusterhead. The presented clustering algorithm requires three types of messages, our algorithms adapted to message passing model require one type of message. A self-stabilizing algorithm for cluster formation is presented in [17]. A density criteria (defined in [18]) is used to select clusterhead: a node v chooses in its neighborhood the node having the highest density. A v's neighborhood contains all nodes at distance less or equal to 2 from v. Therefore, to choose clusterhead, communication at distance 2 is required. Our algorithms build clusters on local information; thus it requires only communication between nodes at distance 1 of each others.

2   Model

In this paper, we consider the state model [19, 20, 21]. A distributed system $\mathcal{S}$ is a set of state machines called processors. Each processor can communicate with a subset of other processors called neighbors. We model a distributed system by an undirected graph $G = (V, E)$ in which V, $|V| = n$, is the set of nodes and there is an edge $\{u, v\} \in E$ if and only if u and v can mutually receive each others' transmission (this implies that all the links between the nodes are bidirectional). In this case we say that u and v are neighbors. The set of neighbors of a node $v \in V$ will be denoted by N_v. Every node v in the network is assigned an unique identifier (ID). For simplicity, here we identify each node with its ID and we denote both with v. We assume the locally shared memory model of communication. Thus, each processor i has a finite set of *local variables* such that the variables at a processor i can be read by i and any neighbors of i, but can only be modified by i. Each processor has a program and the processors execute their programs asynchronously. We assume that the program of each processor i consists of a finite set of guarded statements of the form $Rule : Guard \rightarrow Action$, where $Guard$ is a boolean predicate involving the local variables of i and the local variables of its neighbors, and $Action$ is an assignment that modifies the local variables in i. The *rule* R is executed only if the corresponding guard $Guard$ evaluates to true, in which case we say rule $Rule$ is enabled. The *state* of a processor is defined by the values of its local variables. A *configuration* of a distributed system G is an instance of the processor states. The set of configurations of G is denoted as $\mathcal{C}$. A computation e of a system G is a sequence of configurations $c_1, c_2, \dots$ such that for $i = 1, 2, \dots$, the configuration c_{i+1} is reached from c_i by a single step of one or several processors. A computation is *fair* if any processor in G that is continuously enabled along the computation, will eventually perform an action. *Maximality* means that the computation is either infinite, or it is finite and in this later case no action of G is enabled in the final configuration. Let $\mathcal{C}$ be the set of possible configurations and $\mathcal{E}$ be the set of all possible computations of a system G. The set of computations of G starting with the particular *initial configuration* $c \in \mathcal{C}$ will be denoted $\mathcal{E}_c$. The set of computations of $\mathcal{E}$ whose initial configurations are all elements of $B \in \mathcal{C}$ is denoted as $\mathcal{E}_B$.

In this paper, we use the notion *attractor* [22] to define self-stabilization. Intuitively, an attractor is a set of configurations of a system G that "attracts" a computation of G.

Definition 1. *(Attractor). Let B_1 and B_2 be subsets of $\mathcal{C}$. Then B_1 is an attractor for B_2 if and only if:*

    1. $\forall e \in \mathcal{E}_{B_2}, (e = c_1, c_2, \dots), \exists i \geq 1 : c_i \in B_1$ *(convergence).*

    2. $\forall e \in \mathcal{E}_{B_1}, (e = c_1, c_2, \dots), \forall i \geq 1, c_i \in B_1$ *(closure).*

The set of configurations that matches the specification of problems is called the set of *legitimate* configurations, denoted as $\mathcal{L}$. $\mathcal{C} \backslash \mathcal{L}$ denotes the set of illegitimate configurations.

Definition 2. *(Self-stabilization). A distributed system S is called self-stabilizing if and only if there exists a non-empty set $\mathcal{L} \subseteq \mathcal{C}$ such that the following conditions hold:*

    1. $\mathcal{L}$ *is an attractor for $\mathcal{C}$.*

    2. $\forall e \in \mathcal{E}_{\mathcal{L}}, e$ *verifies the specification problem.*

3 Self-stabilizing DMAC Algorithm

Clustering an ad hoc sensor network means partitioning its nodes into *clusters*, each one with a *clusterhead* and (possibly) some *ordinary nodes*. In order to meet the requirements imposed by the wireless, mobile nature of these networks, ordinary nodes has to be at distance 1 of their clusterhead. Thus, the following *clustering properties* have to be satisfied:

1. Every ordinary node has at least a clusterhead as neighbor (*dominance* property).

2. A clusterhead has not clusterhead neighbors (*independence* property). We consider weighted networks, i.e., a weight w_v is assigned to each node $v \in V$ of the network. In ad hoc sensor networks, amount of bandwidth, memory space or battery power of a processor could be used to determine weight values. For simplicity, in this paper we assume that each node has a different weight.

The choice of the clusterheads will be based on the *weight* associated to each node: the higher the weight of a node, the better this node is suitable to be a clusterhead. Thus the following property has also to be verified:

3. Every ordinary node affiliates with the neighboring clusterhead that has the highest weight.

3.1 Self-stabilizing DMAC Algorithm Description

In the rest of this paper, we will refer to the guard of statement of process v as $G_i(v)$ and the rule of statement of process v as $R_i(v)$.

Constants
    $w_v : \mathbb{N}$; // the weight of node v
Local variables of node v
    Ch_v: boolean; // indicate that v is or is not a clusterhead
    $Clusterhead_v$: IDs // the clusterhead of node v
Predicates
    $G_1(v) \equiv (\forall z \in N_v : (Ch_z = F) \vee (w_v > w_z))$;
    $G_2(v) \equiv (Ch_v = F) \vee (Clusterhead_v \neq v)$;
    $G_3(v) \equiv (Ch_v = T) \vee (Clusterhead_v \neq max_{w_z}\{z \in N_v : Ch_z = T\})$;
Rules
    $R_1(v) : G_1(v) \wedge G_2(v) \rightarrow Ch_v := T; Clusterhead_v := v$;
    $R_2(v) : \neg G_1(v) \wedge G_3(v) \rightarrow Ch_v := F; Clusterhead_v := max_{w_z}\{z : Ch_z = T\}$;

Algorithm 1. Self-stabilizing DMAC algorithm

v is a clusterhead iff $Ch_v = T$ otherwise v is an ordinary node. If v has not a clusterhead in its neighborhood whose weight is higher than its weight then v will become a clusterhead by performing the rule $R_1(v)$. Otherwise, v will be an ordinary node by performing the rule $R_2(v)$; v chooses its clusterhead by selecting the node having the highest weight among its neighbors which are clusterhead. If $Clusterhead$ value on v is not correct according to v'status (i.e., clusterhead or ordinary) then a rule is enabled. The rule $R_1(v)$ (resp. $R_2(v)$.) is enabled if v is a clusterhead (resp. if v is not a clusterhead).

3.2   Proof of Self-stabilizing DMAC Algorithm

Denote $Decided_i$, $i \in \mathbb{N}$ a set of nodes which have certainly selected the cluster-head at end of its step and this clusterhead stays unchange. The convergence is done in step. During the ith step, $\forall p \in Decided_i$ will choose their clusterhead. We define $Decided_i, i \in \mathbb{N}$ as the following recursive rule.

    1: $Decided_0 = \emptyset$.
    2: Denote v_{H_i} the node with the highest weight in $V - Decided_i$.
        $Decided_{i+1} = Decided_i \cup \{v_{H_i} + N_{v_{H_i}} \}$.

We denote $L_i, L'_i, i \in \mathbb{N}$ a set of predicates on processor state. $L'_0 = True$. We will prove that at the end of ith step, L'_{i+1} is verified.

Lemma 3. *Once L'_i is reached, then v_{H_i} becomes a clusterhead and stays forever a clusterhead.*
*($L_{i+1} \equiv L'_i$ **and** $\{v_{H_i}$ is a clusterhead $\}$) is an attractor.*

Lemma 4. *Once L_{i+1} is reached, all v_{H_i}'s neighbors in V_i choose v_{H_i} as their clusterhead and keep it.*
*($L'_{i+1} \equiv L_i$ **and** $\{\forall u \in (N_{v_{H_i}} \cap V_i) : Clusterhead_u = v_{H_i} \}$) is an attractor.*

Theorem 5. *The system eventually reaches a terminal configuration.*

Theorem 6. *Once the terminal configuration is reached, the clustering proper-ties are satisfied.*

Detailed proofs of convergence and correctness of Self-stabilizing DMAC algorithm can be found in [23].

4   Self-stabilizing GDMAC Algorithm

In the previous algorithm, we have requirement that the clusterheads are bound to never be neighbors. This implies that, when due to the mobility of the processors two or more clusterheads become neighbors, those with the smaller weights have to *resign* and affiliate with the now higher neighboring clusterhead. Furthermore, when a clusterhead v becomes the neighbor of an ordinary processor u whose current clusterhead has weight smaller than v's weight, u has to affiliate with (i.e., *switch* to the cluster of) v. These "resignation" and "switching" processes due to processor's mobility are a consistent part of the clustering management overhead that should be minimized in ad hoc sensor networks. To overcome the above limitations, we introduce in this section a generalization of the previous algorithm. This algorithm is used to partition the nodes of the networks so that the following three *ad hoc sensor clustering* properties are satisfied.

    1. Every ordinary node always affiliates with (only) one clusterhead which has higher weight than its weight (*affiliation* condition).

    2. For every ordinary node v, for every clusterhead $z \in N_v$:
    $w_z \leq w_{Clusterhead_v} + h$ (*clusterhead* condition).

3. A clusterhead has at most k neighboring clusterheads (k being an integer, $0 \leq k < n$) (*k-neighborhood* condition).

The first requirement ensures that each ordinary node has direct access to at least one clusterhead (the one of the cluster to which it belongs), thus allowing fast intra and inter cluster communications. The second requirement guarantees that each ordinary node always stays with a clusterhead that gives it a "good" service. By varying the threshold parameter h it is possible to reduce the switching overhead associated to the passage of an ordinary node from its current clusterhead to a new neighboring one when it is not necessary. With this requirement we want to incur the switching overhead only when it is really convenient. When $h = 0$ we simply obtain that each ordinary node affiliates with the neighboring clusterhead with the highest weight. Finally, the third requirement allows us to have up to k neighboring clusterheads, $0 \leq k < n$. When $k = 0$ we obtain that two clusterhead can not be neighbors. Notice that the case with $k = h = 0$ corresponds to the previous algorithm.

Constants

$\quad w_v : \mathbb{N};$ // the weight of node v

Local variables of node v

$\quad Ch_v$: boolean; // indicate that v is or is not a clusterhead.

$\quad Clusterhead_v : IDs$ // the clusterhead of node v.

$\quad SR_v : \mathbb{N}$ // the highest weight which violates the 3th condition in v's neighbor.

Macros

$\quad N_v^+ = \{z \in N_v : (Ch_z = T) \wedge (w_z > w_v)\};$ // the set of v's neighboring clusterhead which has higher weight than v's weight.

$\quad Cl_v = |N_v^+|;$ // the number of v's neighboring clusterhead which has higher weight than v's weight.

Predicates

$\quad \mathbf{G_1}(v) = \mathbf{G_{11}}(v) \vee \mathbf{G_{12}}(v)$

$\quad \mathbf{G_{11}}(v) \equiv [(Ch_v = F) \wedge (N_v^+ = \emptyset)]$

$\quad \mathbf{G_{12}}(v) \equiv [(Ch_v = T) \wedge (Clusterhead_v \neq v) \wedge (\forall z \in N_v^+ : w_v > SR_z) \wedge (Cl_v \leq k)]$

$\quad \mathbf{G_2}(v) = \mathbf{G_{21}}(v) \vee \mathbf{G_{22}}(v)$

$\quad \mathbf{G_{21}}(v) \equiv [(Ch_v = F) \wedge \{(\exists z \in N_v^+ : w_z > w_{Clusterhead_v} + h)$
$\qquad\qquad\qquad\qquad\qquad\qquad\qquad\qquad\qquad\qquad \vee (Clusterhead_v \notin N_v^+)\}]$

$\quad \mathbf{G_{22}}(v) \equiv [(Ch_v = T) \wedge \{(\exists z \in N_v^+ : (w_v \leq SR_z)) \vee (Cl_v > k)\}]$

$\quad \mathbf{G_3}(v) \equiv (Ch_v = F) \wedge (SR_v \neq 0)$

$\quad \mathbf{G_4}(v) \equiv (Ch_v = T) \wedge (SR_v \neq max(0, k + 1^{\text{th}}\{w_z : z \in N_v \wedge (Ch_z = T)\}))$

Rules

$\quad \mathbf{R_1}(v) : \mathbf{G_1}(v) \rightarrow Ch_v := T; Clusterhead_v := v;$
$\qquad\qquad\qquad\qquad\qquad SR_v := max(0, k + 1^{\text{th}}\{w_z : z \in N_v \wedge (Ch_z = T)\});$

$\quad \mathbf{R_2}(v) : \mathbf{G_2}(v) \rightarrow Ch_v := F; Clusterhead_v := max_{w_v}\{z \in N_v^+\}; SR_v := 0;$

$\quad$ // update the value of SR_v

$\quad \mathbf{R_3}(v) : \mathbf{G_3}(v) \rightarrow SR_v := 0;$

$\quad \mathbf{R_4}(v) : \mathbf{G_4}(v) \rightarrow SR_v := max(0, k + 1^{\text{th}}\{w_z : z \in N_v \wedge (Ch_z = T)\});$

Algorithm 2. Self-stabilizing GDMAC algorithm

4.1   GDMAC Algorithm Description

Similarly to the algorithm 1, the rule R_1 sets up the performing node as a clusterhead; after the R_2 action, the performing node is ordinary. A clusterhead v checks the number of its neighbors that are clusterheads. If they exceed k, then it sets up the value of SR_v to the weight of the first clusterhead (namely, the one with the $(k+1)$th highest weight) that violates the k-neighborhood condition (R_4 action). Otherwise, SR_v is assigned to 0 (R_3 action). SR_v value of an ordinary node is 0 or R_3 is enabled to set the value to 0.

We split the possibles cases where a node v has to change its role (i.e., to become ordinary or clusterhead) in the following mutually exclusive ones:

Case 1. v is an ordinary node and v cannot select any neighbors as clusterhead - otherwise the affiliation condition will be violated. $G_{11}(v)$ is verified: v will become a clusterhead (R_1 action).

Case 2. v is a clusterhead. v does not violate the k-neighborhood condition but the value v's clusterhead is incorrect. $G_{12}(v)$ is verified: v will correct the value of its clusterhead (R_1 action).

Case 3. v is an ordinary node and v violates the *clusterhead* condition. $G_{21}(v)$ is verified: v will become an ordinary node (R_2 action).

Case 4. v is a clusterhead and v violates the k-neighborhood condition. $G_{22}(v)$ is verified: v will become an ordinary node (R_2 action).

4.2   Proof of Convergence

We first prove that the system reaches a terminal configuration.

Lemma 7. $A_1 = \{\mathcal{C} \mid \forall v : G_{12}(v) = F\}$ *is an attractor.*

Proof. If v verifies predicate $G_{12}(v)$ then v is enabled and will stay enabled up to the time where v performs $R_1(v)$. As all computations are fair, v eventually performs $R_1(v)$. After that $G_{12}(v)$ is never verified (see the rule action).   □

Lemma 8. *In A_1, once v had performed a rule $R_1(v)$ or $R_2(v)$, the guard of statements $G_i(v) : i = 1, 2$ remain false unless there exists a node u, $w_u > w_v$, that performs a rule $R_1(u)$ or $R_2(u)$.*

Proof. In A_1, $G_{12}(v)$ is never true.

Case 1. Once v had performed the rule $R_1(v)$, we have that $Ch_v = T$ and $Clusterhead_v = v$. Thus, the next rule performed by v will be $R_2(v)$.
Before doing $R_1(v)$, $G_{11}(v)$ is verified, we have $N_v^+ = \emptyset$. At time where v performs $R_2(v)$, $G_{22}(v)$ is verified, implies that $N_v^+ \neq \emptyset$, thus there is a node $u \in N_v$, $w_u > w_v$ that performed the rule $R_1(u)$ in meantime.

Case 2. Once v had performed the rule $R_2(v)$, we have that $Ch_v = F$ and $Clusterhead_v := max_{w_z}\{z \in N_v : Ch_z = T\}$, next time v would perform a rule only if $G_{11}(v)$ **or** $G_{21}(v)$ is verified. Denote u the clusterhead of v, then after doing $R_2(v)$ we have $u \in N_v^+$ and $w_u = max(w_z, \forall z \in N_v^+)$.

Case 2.1. v will performs $R_1(v)$ because $G_{11}(v)$ is verified. At time where v performs $R_1(v)$, $G_{11}(v)$ is verified then $N_v^+ = \emptyset$, implies that u performed the rule $R_2(u)$ in meantime.

Case 2.2. v will performs $R_2(v)$ because $G_{21}(v)$ is verified. G_{21} is verified, means that $(\exists z \in N_v^+ : w_z > w_u + h) \vee (u \notin N_v^+)$, implies that there exists a node $z \in N_v, w_z > w_u + h > w_u$ performed $R_1(z)$ **or** u performed $R_2(u)$ in meantime.   □

Lemma 9. $A_2 = A_1 \cup \{C | \forall v : (G_1(v) = F) \wedge (G_2(v) = F)\}$ *is an attractor.*

Proof. We will prove by contradiction. Assume that A_2 is not an attractor. A processor cannot verify forever $G_1 \vee G_2$ (this processor would be enabled forever and never performs a rule). Thus along a maximal computation there is a processor v that infinitely often verifies $G_1(v)$ or $G_2(v)$ and also infinitely often does not verify $G_1(v)$ or $G_2(v)$. Meaning that v executes infinitely often $R_1(v)$ or $R_2(v)$. Following Lemma 4, once v have performed a rule $R_1(v)$ or, $R_2(v)$ it would perform $R_1(v)$ or $R_2(v)$ again if there exists a processor u $(w_u > w_v)$ that performs $R_1(u)$ or $R_2(u)$. Since the set of processors is finite, then v performs $R_1(v)$ or $R_2(v)$ infinitely often only if there exists a processor u $(w_u > w_v)$ that performs $R_1(u)$ or $R_2(u)$ infinitely many times. Using a similar argument we have an infinite sequence of processors having increasing weight that performs R_1 or R_2 infinitely often. Since the number of processors is finite, this is a contrary. Hence our hypothesis is false, and for every node v, $G_i(v) : i = 1, 2$ becomes eventually false and stay false.   □

Theorem 10. *The system eventually reaches a terminal configuration.*

Proof. By Lemma 5, A_2 is an attractor. In A_2, processor v would only update of SR_v one time if necessary.   □

4.3   Proof of Correctness

Theorem 11. *Once a terminal configuration is reached, the ad hoc sensor clustering properties are satisfied.*

Proof. In a terminal configuration, for every processor v, we have $G_i(v) = F :$ $i = 1, 2$.

Case 1. v is an ordinary node.
$G_1(v) = F$ implies N_v^+ is not empty. $G_2(v) = F$ implies $(\nexists z \in N_v^+ : (w_z > w_{Clusterhead_v} + h))$ and $(Clusterhead_v \in N_v^+)$. Thus v satisfies property 1 and 2.

Case 2. v is a clusterhead node.
$(G_2(v) = F) \equiv (\forall z \in N_v^+ : w_v > SR_z) \wedge (Cl_v \leq k)$. $G_1(v) = F$ implies that $Clusterhead_v = v$. We now prove that v has at most k neighboring clusterheads. Since $Cl_v \leq k$, then v has at most k neighboring clusterheads with higher weight

than v's weight. Assume that v has more than k neighboring clusterheads, thus there exits at least a neighboring clusterhead u of v such that $w_u \leq SR_v < w_v$. Hence, $G_{22}(u) = T$ because $v \in N_u^+(w_u \leq SR_v)$, that is a contrary. $\square$

5   Time Complexity

The stabilization time is the maximum number of rounds needed to reach a stabilized state from an arbitrary initial one. Figure 1 presents a scenario to measure stabilization time of GDMAC in the case $k = 1$, $h = 0$. Notice that the scenario can be easily generalized at any value of k and the initial configuration is the worst one. We have a system S composed by m blocs as depicted in Figure 1(a). Each bloc B_i includes two clusterheads X_i, Y_i and an ordinary node Z_i. We assume that the weight of nodes are ordered as the following: $X_i > Y_i > Z_i > Y_{i+1}$. A clusterhead node Z', $Z' > Y_1$ is a neighbor of Y_1. The largest convergence time under any weight-based clustering algorithms happens with this initial configuration. We denote N the number of nodes in the system S, $N = m(k+2)+1$. Following Algorithm 2, each bloc B_i will one after another takes two rounds to reconstruct under the synchronous schedule. Thus, $2m$ rounds are needed to converge under the synchronous schedule. Then, the stabilization time is $O(2N/(k+2))$ in this example. Notice that for any k value, there is a network topology and an initial configuration such that the convergence time is $O(D)$, where D is network diameter.

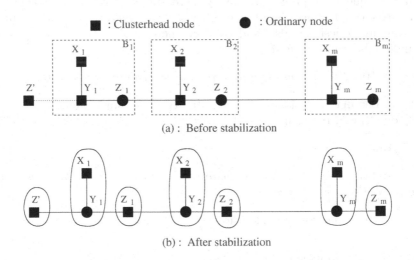

(a) : Before stabilization

(b) : After stabilization

Fig. 1. Stabilization time

6   Concluding Remarks

The presented algorithms are designed for the state model. Nevertheless, our algorithms can be easily transformed into algorithms for the message-passing model. Each node v periodically broadcasts to its neighbors its state (i.e., a message containing the values of Ch_v, SR_v, and w_v). Based on this message, v's neighbors decide to update or not their variables. After a change in the value of Ch_v, SR_v, or w_v, a node v broadcasts to its neighbors its new state.

We have computed the following upper bound of the stabilization time of self-stabilizing $GDMAC$ in a case where the network diameter is $N/(k+2)$: $O(2N/(k+2))$. When $k = 0$, this bound is $O(D)$. Using DMAC algorithm, the convergence is also $O(D)$. Any self-stabilizing algorithm building clusters that verify the *ad hoc sensor clustering* properties has a stabilization time of $O(D)$ on some network topologies. By varying the threshold parameter k, the stabilization time can be reduced. Reducing the stabilization time is very important for ad hoc sensor networks: topology changes happen fairly often, conservation of sensor energy is a key factor. Thus, network management procedure should be as simple as possible.

The presented algorithms require bidirectional link. In wireless sensor networks, some of the links are unidirectional. A node can send data to another node, but this latter cannot reply (more precisely, its reply will not reach its destination). Extend our algorithms to unidirectional links is left for the future work.

References

[1] Banerjee, S., Khuller, S.: A clustering scheme for hierarchical control in multi-hop wireless networks. In: INFOCOM'01: The 20th Conference of the IEEE Communications Society. (2001) 1028–1037

[2] Basagni, S.: Distributed and mobility-adaptive clustering for multimedia support in multi-hop wireless networks. In: VTC'99: Proceedings of the IEEE 50th International Vehicular Technology Conference. (1999) 889–893

[3] Basagni, S.: Distributed clustering for ad hoc networks. In: ISPAN'99: Proceedings of the 1999 International Symposium on Parallel Architectures, Algorithms, and Networks. (1999) 310–315

[4] Chatterjee, M., Das, S., Turgut, D.: WCA: A weighted clustering algorithm for mobile ad hoc networks. Journal of Cluster Computing, Special issue on Mobile Ad hoc Networking 5 (2002) 193–204

[5] Fernandess, Y., Malkhi, D.: K-clustering in wireless ad hoc networks. In: POMC'02: Proceedings of the second ACM international workshop on Principles of mobile computing. (2002) 31–37

[6] Gerla, M., Tsai, J.T.: Multicluster, mobile, multimedia radio network. Wireless Networks 1 (1995) 255–265

[7] Lin, C.R., Gerla, M.: Adaptive clustering for mobile wireless networks. IEEE Journal on Selected Areas in Communications 15 (1997) 1265–1275

[8] Bettstetter, C., Friedrich, B.: Time and message complexities of the generalized distributed mobility-adaptive clustering (GDMAC) algorithm in wireless multihop networks. In: VTC'03: Proceedings IEEE Vehicular Technology Conference. (2003) 176–180

[9] Bettstetter, C., Krausser, R.: Scenario-based stability analysis of the distributed mobility-adaptive clustering (DMAC) algorithm. In: MobiHoc'01: Proceedings of the 2nd ACM Symposium on Mobile Ad Hoc Networking & Computing. (2001) 232–241

[10] Heinzelman, W., Chandrakasan, A., Balakrishnan, H.: An application-specific protocol architecture for wireless microsensor networks. IEEE Transactions on Wireless communications 1 (2002) 660–670

[11] Younis, O., Fahmy, S.: Distributed clustering for ad-hoc sensor networks: A hybrid, energy-efficient approach. In: IEEE INFOCOM'04: Proceedings of The 23rd Conference of the IEEE Communications Society. (2004)

[12] Dolev, S.: Self-Stabilization. MIT Press (2000)

[13] Schneider, M.: Self-stabilization. ACM Symposium Computing Surveys 25 (1993) 45–67

[14] Goddard, W., Hedetniemi, S.T., Jacobs, D.P., Srimani, P.K.: Self-stabilizing protocols for maximal matching and maximal independent sets for ad hoc networks. In: WAPDCM'03: 5th IPDPS Workshop on Advances in Parallel and Distributed Computational Models. (2003)

[15] Xu, Z., Hedetniemi, S.T., Goddard, W., Srimani, P.K.: A synchronous self-stabilizing minimal domination protocol in an arbitrary network graph. In: IWDC'03: Proceedings of the 5th International Workshop on Distributed Computing, Springer LNCS 2918. (2003)

[16] Bein, D., Datta, A.K., Jagganagari, C.R., Villain, V.: A self-stabilizing link-cluster algorithm in mobile ad hoc networks. In: ISPAN'05: Proceedings of the 8th International Symposium on Parallel Architectures, Algorithms and Networks. (2005) 436–441

[17] Mitton, N., Fleury, E., Lassous, I.G., Tixeuil, S.: Self-stabilization in self-organized multihop wireless networks. In: WWAN'05: Proceedings of the 25th IEEE International Conference on Distributed Computing Systems Workshops. (2005) 909–915

[18] Mitton, N., Busson, A., Fleury, E.: Self-organization in large scale ad hoc networks. In: In MED-HOC-NET'04: Third Annual Mediterranean Ad Hoc Networking Workshop. (2004)

[19] Beauquier, J., Gradinariu, M., Johnen, C.: Memory space requirements for self-stabilizing leader election protocols. In: PODC'99: Proceedings of the eighteenth annual ACM symposium on Principles of distributed computing. (1999) 199–207

[20] Johnen, C.: Service time optimal self-stabilizing token circulation protocol on anonymous unidirectional rings. In: SRDS'02: Proceedings of the 21st IEEE Symposium on Reliable Distributed Systems. (2002) 80–89

[21] Johnen, C., Alima, L.O., Tixeuil, S., Datta, A.K.: Self-stabilizing neighborhood synchronizer in tree networks. In: ICDCS'99: Proceedings of the 19th IEEE International Conference on Distributed Computing Systems. (1999) 487

[22] Johnen, C., Tixeuil, S.: Route preserving stabilization. In: SSS'03: Proceedings of the 6th International Symposium on Self-stabilizing System, Springer LNCS 2704. (2003) 184–198

[23] Johnen, C., Nguyen, L.: Self-stabilizing clustering algorithm for ad hoc networks. Technical Report no. 1357, L.R.I, Université de Paris Sud 1429 (2006)

Improved Stretch Factor for Bounded-Degree Planar Power Spanners of Wireless Ad-Hoc Networks[*]

Iyad A. Kanj[1] and Ljubomir Perković[2]

[1] School of Computer Science, Telecommunications and Information Systems, DePaul University, 243 S. Wabash Avenue, Chicago, IL 60604-2301, USA
ikanj@cs.depaul.edu
[2] School of Computer Science, Telecommunications and Information Systems, DePaul University, 243 S. Wabash Avenue, Chicago, IL 60604-2301, USA
lperkovic@cs.depaul.edu

Abstract. Given a wireless Ad-Hoc network modeled as a unit disk graph U in the plane, we present a localized distributed algorithm that constructs a bounded degree planar power spanner of U with a bounded stretch factor. More specifically, for an integer parameter $k \geq 8$ and a power exponent constant $p \in [2, 5]$, our algorithm constructs a planar power spanner for the network of degree bounded by $k + 5$ and a stretch factor bounded by $1 + 2^p \sin^p(\pi/k)$. This significantly improves the previous best results in the literature by Song et al..

1 Introduction

A wireless ad-hoc network is a decentralized and self-configuring network of mobile devices connected using wireless channels. The wireless ad-hoc network is usually modeled as a *unit disk graph* in the two dimensional plane, whose points correspond to wireless devices and whose edges connect pairs of points whose corresponding devices are in each other's transmission range. All devices are assumed to have the same transmission range equal to one unit. Associated with each edge is the cost to support the corresponding link in the network, which is assumed to be the Euclidian distance between the endpoints raised to some exponent p between 2 and 5. Each point is responsible for constructing and maintaining a local topology with a subset of neighboring points. Distant points then communicate through messages relayed by intermediate neighbors. The cost of distant points communicating is the sum of the costs of the edges of the path formed by the intermediate points. The shortest path between any pair of points will be the path connecting the pair of points that has the smallest energy cost. Energy consumption is a critical issue for (battery-powered) mobile devices, and the primary goal when constructing a backbone topology for wireless ad-hoc networks is to make it energy efficient.

---

[*] The first author was supported in part by DePaul University Competitive Research Grant.

S. Nikoletseas and J.D.P. Rolim (Eds.): ALGOSENSORS 2006, LNCS 4240, pp. 95–106, 2006.

Because of interference and contention issues, a major constraint on the network topology is the requirement that each device maintains links to only a constant number of devices in its transmission range. This will also allow the devices to decrease their transmission power to levels required to reach the selected devices only. The network topology should also be amenable to guaranteed and efficient routing. The folklore "right hand rule" (in face routing), discussed in [1], is one of a few routing rules that require the network to be planar. Finally, the construction and maintenance of the network topology should be done in a localized distributed manner, where devices construct and maintain links based only on the information from neighboring devices. The number of messages and the size of the messages required to construct and maintain the topology should be low. The problem of computing efficient topologies for wireless ad-hoc networks was extensively studied in various settings (see for instance [8,9,3,5,6,4,7,10]).

The corresponding topology control problem consists therefore of finding a low cost, constant degree planar subgraph of the unit disk graph using a distributed localized algorithm. This subgraph is referred to as the power spanner, and the low cost requirement can be quantified as follows: a subgraph is a power spanner with stretch factor ρ if the cost of the shortest path in the subgraph between any pair of points is at most ρ times the cost of the shortest path in the graph itself.

Song, Wang, Li, and Frieder proposed algorithms [8] for this problem with proven bounds on the maximum degree and stretch factors. Given a parameter $k > 6$, their first algorithm computes a power spanner of maximum degree $k + 5$ and maximum stretch factor $\rho = 1/(1 - (2 \sin(\pi/k))^p)$. In their second algorithm, given a parameter $k > 8$, they obtain a bound of k for the maximum degree and $\rho = (\sqrt{2})^p/(1 - (2\sqrt{2} \sin(\pi/k))^p)$ for the stretch factor. Both algorithms use the well-known notions of Gabriel and Yao subgraphs.

In this paper we present an algorithm that significantly improves the results in [8]. Given an integer parameter $k > 7$, our algorithm's bounds for the maximum degree and stretch factor are $k + 5$ and $\rho = 1 + 2^p \sin^p(\pi/k)$, respectively. This significantly improves the stretch factors in [8] for the same degree bound, as shown in Table 1 in Section 6.

The paper is organized as follows. Section 2 reviews the related definitions and background. In Section 3 we define the notion of a *canonical path*, crucial to the results in this paper. In Section 4 we present a simple algorithm that handles the cases when the parameter $k \geq 12$, and we extend this algorithm in Section 5 to handle the cases when the parameter $k \in \{8, 9, 10, 11\}$. We conclude the paper in Section 6 by comparing our results to the relevant results in the literature. Many proofs are omitted for the lack of space.

2   Background and Preliminaries

A wireless network consists of a set of n points in the two dimensional plane. Each point has a transmission range of one unit; in other words, two points A and B can transmit to each other if their Euclidian distance $|AB|$ is at most

1 unit. It is assumed that each point knows its coordinates through a Global Position System (GPS). A *unit disk graph* U is therefore defined on the n points as follows: for every two points A and B, AB is an edge in U if and only if $|AB| \leq 1$. The edge AB is embedded in the plane as the straight line segment AB. The unit disk graph U will be assumed connected. The power required to support a link/edge AB in U is commonly assumed to be $|AB|^p$, where p is a constant in the interval $[2, 5]$. Two far apart points A and B communicate through intermediate points that form a simple path $A = M_0, M_1, ..., M_r = B$ in U. The energy cost of this path is:

$$\sum_{j=0}^{r-1} |M_j M_{j+1}|^p.$$

Among all paths between A and B, a path in U with the smallest energy cost is defined to be a *smallest cost path* and we denote its cost as $c_U(A, B)$. A subgraph H of U is a *power spanner* if there is a constant ρ such that for every two points $A, B \in U$ we have: $c_H(A, B) \leq \rho c_U(A, B)$. The constant ρ is called the *stretch factor* of H. The following lemma is from [8]:

Lemma 1 ([8]). *A subgraph H of graph U has stretch factor ρ if and only if for every edge $AB \in U$, $c_H(A, B) \leq \rho c_U(A, B) = \rho |AB|^p$.*

In this paper we present algorithms that construct a bounded degree planar power spanner of U with a very small power stretch factor ρ. Most of the previous work on bounded degree planar power spanners is based on the concepts of Gabriel and Yao subgraphs, which are essential for our current work. We review these concepts next.

The *Gabriel subgraph* G of a unit disk graph U is obtained by removing every edge $AB \in U$ such that there is a point $M \in U$ with $|MA|^2 + |MB|^2 \leq |AB|^2$, i.e., M is contained in the closed disk of diameter AB ([2]). The following properties were shown in [8].

Proposition 1 ([8]). *Let U be a unit disk graph and let G be the Gabriel subgraph of U.*

1. *G is planar.*
2. *G is connected.*
3. *The power stretch factor of G is 1.*
4. *If AB is an edge in U, then $AB \in G$ if and only if for every point M in U the angle $\angle AMB$ in the interior of triangle $\triangle AMB$ is acute.*

The Yao subgraph [11] of a planar graph embedded on the plane with integer parameter $k > 6$, is constructed by repeating the following step, which we will call a *Yao step*, for every point M: k *equally separated rays out of M are arbitrarily defined, and k closed cones of size $2\pi/k$ are thus created; then, in each cone, the shortest edge MN inside the cone (if any) is chosen and added to the Yao subgraph.*

Song et al. [8] applied a Yao subgraph construction to a Gabriel graph G to obtain G', a bounded degree planar power spanner of U. They noted that applying the Yao construction directly to G does not lead to a bounded degree subgraph. Instead, they first oriented the edges of the Gabriel graph G (using the classical acyclic orientation of a planar graph described in more detail later) so that every point in G has in-degree at most 5. Then, they applied the above Yao step to every point of G but to the *outgoing edges only*. The subgraph G' thus obtained has then maximum degree $k + 5$.

When an outgoing edge CB (from C) of a directed Gabriel graph G is not included in the Yao subgraph G', there must exist an outgoing edge $CA \in G$ such that: $|CA| \leq |CB|$, $\angle BAC = \alpha \leq 2\pi/k$, and $|AB| \leq 2\sin(\pi/k)|CB|$, as shown in [8]. The edge CA is then used in [8] as the first edge of a path in G' from C to B. If $AB \in G'$, then this path in G' consists of the edges CA and AB, and has cost at most $(1 + 2^p \sin^p(\pi/k))|CB|^p$. If, however, $AB \notin G'$, then a path from A to B is recursively shown to exist, and its total cost is bounded by $\rho|CB|^p$, where $\rho \leq 1/(1 - (2\sin(\pi/k))^p)$. Therefore, a planar power spanner G' of U with a bounded degree $k + 5$ and a stretch factor bounded by $1/(1 - (2\sin(\pi/k))^p)$ was given in [8].

In this paper, we modify the Yao step so that a low cost path from A to B in G is always included in the Yao subgraph G'. In fact, this path, which we will call the *canonical path*, has cost at most $|AB|^p$, giving us a stretch factor of $1 + 2^p \sin^p(\pi/k)$—significantly smaller than the stretch factor of [8], for the same degree bound of $k + 5$.

3   Canonical Paths

We assume that G is the Gabriel subgraph of U. Let A and B be two points in U and suppose that $AB \in U$. If $AB \notin G$, then by Proposition 1, there must exist at least one point $M \in U$ such that $\angle AMB \geq \pi/2$. We define the *canonical point* for the pair (A, B) as the point M with the *largest* such angle $\angle AMB$.

Proposition 2. *Let A and B be two points in U such that the edge $AB \in U$ but $AB \notin G$, and let M be the canonical point for the pair (A, B). Then the following are true.*

1. *There is no point of U inside the triangle $\triangle AMB$.*
2. *Both MA and MB are edges in U.*
3. *For any $p \in [2, 5]$, $|MA|^p + |MB|^p \leq |AB|^p$.*
4. *If there is a point $C \in U$ such that CA and CB are edges in G and $\angle BCA$ is acute, then the canonical point M for (A, B) is inside the half-plane not containing the point C determined by the straight line AB.*

Let CA and CB be edges in G. We now define the *canonical path* from $X = A$ to $Y = B$ to be the path in G constructed recursively as follows: *if $XY \in G$ then edge XY is returned, otherwise the concatenation of the canonical paths from X to M and from M to Y is returned, where M is the canonical point for (X, Y).*

Let $F(C, A, B)$ be the region in G bordered by edges CA, CB, and the edges of the canonical path from A to B. If M is a canonical path point, we define the *angle at* M to be the angle inside the region $F(A, B, C)$ and formed by edges of the canonical path incident to M. The following theorem describes the structure of canonical paths.

Theorem 1. *Let CA and CB be edges in G such that $\angle BCA = \alpha \leq \pi/4$. Then the following are true. (See Figure 1.)*

1. *The interior of $F(C, A, B)$ may contain only edges connecting C to points of the canonical path.*
2. *If $|CA| \leq |CB|$ then the angle $\angle CAB$ satisfies $\angle CAB \geq (\pi - \alpha)/2$ and the angle $\angle ABC$ satisfies $\angle ABC \geq \pi/2 - \alpha$.*
3. *Each angle at a canonical path point is at least $\pi/2$.*
4. *If there is an edge $CM \in G$ inside $F(C, A, B)$ from C to a point M on the canonical path such that CM is not exterior to $F(C, A, B)$, then the angle at M is at least $3\pi/4$, and if L and N are the points adjacent to M on the canonical path, then each of the angles $\angle CML$ and $\angle CMN$ is at least $\pi/4$.*
5. *The cost of the canonical path from A to B is at most $|AB|^p$.*

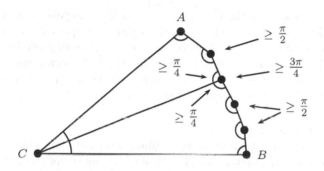

Fig. 1. The general structure of the canonical path

We use Theorem 1 to define next certain structures that form the canonical path edges. Our algorithms, when computing the subgraph G' of G, will guarantee that all edges on the canonical paths are added to G' by simply ensuring that edges of those structures are added to G'.

Let M be a point in G and let e and e' be two edges incident to M. The two edges e and e' are said to be *consecutive*, if one of the angular sectors formed by e and e' contains no edge other than e and e'. We define a *sparse sector* $S_{\geq 2}$ around point M to be a sequence of $\ell \geq 2$ consecutive edges incident to M such that the angle determined by the first edge and the last edge in the sequence is at least $\ell\pi/4$ and the angle between any two consecutive edges in the sequence is at least $\pi/4$. Note that, in particular, two consecutive edges forming an angle of at least $\pi/2$ form the sparse sector $S_{\geq 2}$ around M. The sequence of edges in a sector will be referred to as the *sector edges*. A sparse sector around M is *maximal* if the sector edges are not properly contained in the sector edges of any other sparse sector around M. We also define a *sector*, denoted S_1, around a point M of G, to be a sequence of two consecutive edges that form an angle of at least $(k-2)\pi/2k$, where $k \geq 8$ is an integer parameter. Note that the S_1 sector angle will always be at least $3\pi/8$ for any value of $k \geq 8$.

We will next describe two algorithms for constructing bounded degree power spanners G' of G (and therefore of U).

4   A Simple Algorithm

Our first simple algorithm for constructing a bounded degree power spanner works for parameter values $k \geq 12$.

Construction of the Gabriel Subgraph of U
As in [8], we start by constructing the Gabriel subgraph G of U. This step can be done by exchanging no more than $O(m)$ messages–as each point needs to send its coordinates to its neighbors, where m is the number of edges in U.

Orienting G
We continue by orienting the edges of G to make G acyclic and of maximum in-degree 5. This classical orientation of a planar graph can be obtained by repeatedly choosing a point A of degree at most 5, orienting the incident edges into A, and then removing A and its incident edges. This can easily be done in a distributed manner by exchanging no more than $O(n)$ (G is planar) messages as shown in [8], where n is the number of points in U.

The Modified Yao Step
We color the points in G with in-degree 0 black, and the remaining points white. When an unprocessed point A is colored black it performs the following algorithm. Let $Out(A)$ be the set of all outgoing edges from A.

The Algorithm Edge_Selection
(1) perform a standard Yao step on the set of edges $Out(A)$ to define k cones out of A, each forming an angle of $2\pi/k$, and select up to $r \leq k$ outgoing edges (each of these edges is the shortest edge in its respective nonempty cone); note that $k - r \geq 0$ cones will contain no edges in $Out(A)$;

(2) **for** every maximal sparse sector around A **do** select the sector edges in $Out(A)$;

(3) **for** every unselected edge $AB \in Out(A)$ such that there exists a selected incoming (to A) consecutive edge CA with $\angle CAB \geq (k-2)\pi/2k$ and such that CA is not a sector edge in any maximal sparse sector around A **do** select AB;

(4) **for** every unselected edge $AM \in Out(A)$ such that there exists an unselected consecutive edge NA incoming to A with $\angle NAM \geq (k-4)\pi/2k$ **do** select AM;

(5) set the status of every unselected edge in $Out(A)$ to unselected, send a message to every outgoing neighbor B (i.e., AB is an outgoing edge) notifying it whether the edge AB has been selected or not, and set the status of A to processed.

Upon receiving a message from a neighbor B, point A performs the following steps:

(1') update the status of the edge BA according to the message received from B;

(2') **if** for every incoming neighbor B (i.e., BA is an incoming edge) the status of the edge BA has been determined **then** color(A):= black.

Theorem 2. *Let CA and CB be oriented edges in G such that $|CA| \leq |CB|$ and $\angle BAC = \alpha \leq 2\pi/k$, where $k \geq 8$. Suppose that $CA \in G'$ and $CB \notin G'$. Then every edge of the canonical path between A and B is in G'.*

Proof. Let $(A = V_0, \cdots, V_s = B)$ be the canonical path between A and B. Note that, by part 2 of Theorem 1, $\angle CAB \geq (\pi - \alpha)/2 \geq (k-2)\pi/2k$, and $\angle CBA \geq \pi/2 - \alpha \geq (k-4)\pi/2k$. Since $\angle CAV_1 \geq \angle CAB$ and $\angle CBV_{s-1} \geq \angle CBA$, we have $\angle CAV_1 \geq (k-2)\pi/2k$ and $\angle CBV_{s-1} \geq (k-4)\pi/2k$. It follows that the edge AV_1 will be selected by step (3) of the algorithm if it is oriented from A to V_1, and the edge BV_{s-1} will be selected by step (4) of the algorithm if it is oriented from B to V_{s-1}. (Note that $V_1 = B$ and $V_{s-1} = A$ if the canonical path between A and B consists of the edge AB.) Now every edge on the canonical path other than AV_1 when it is oriented from A to V_1 and BV_{s-1} when it is oriented from B to V_{s-1}, will be an outgoing edge from some canonical point V_i, $i \notin \{0, s\}$, and will be an outgoing edge in some maximal sparse sector around V_i by parts 3 and 4 of Theorem 1. Therefore, every such edge on the canonical path will be selected by part (2) of the algorithm. This completes the proof.

This proof of the following lemma can be found in Lemma 6 in [8].

Lemma 2 (Lemma 6, [8]). *Let A, B, C be three points in a Gabriel graph G such that CB and CA are edges in G. Suppose that $|CA| \leq |CB|$ and $\angle BCA < \alpha$ for some $\alpha \in (0, \pi/2)$. Let $p \in [2,5]$ be a constant. Then $|CA|^p + |AB|^p \leq (1 + 2^p \sin^p(\alpha/2))|CB|^p$.*

Let G' be the subgraph of G whose point set is the point set of G and whose edges are the edges selected by the algorithm **Edge_Selection**. The algorithm **Edge_Selection** is a localized distributed algorithm that does not exchange more than $O(n)$ messages, and hence G' can be constructed from G by exchanging no more than $O(n)$ messages.

Theorem 3. *The subgraph G' is a power spanner of G of maximum degree $k+5$ and a stretch factor $\rho \leq 1 + 2^p \sin^p(\pi/k)$.*

Proof. Since the power stretch factor of G is 1 by Lemma 1, to prove that the stretch factor of G' is bounded by $1 + 2^p \sin^p(\pi/k)$, it suffices to show by Lemma 1 *that every edge in G has a power stretch factor bounded by $1 + 2^p \sin^p (\pi/k)$ in G'.*

Let CB be an edge in G. If CB is an edge of G' then we are done. Suppose now that CB is not an edge in G'. Then CB was not selected by the algorithm. Without loss of generality, assume that the edge CB was directed from C to B in the oriented graph G (otherwise, exchange C and B). The algorithm guarantees by step (1) (the Yao step) that, for every outgoing edge $CB \in G$ (from C) not included in G', G' will contain an outgoing edge CA (from C) such that $|CA| \leq |CB|$ and $\angle BCA = \alpha \leq 2\pi/k$.

Since CA and CB satisfy the statement of Theorem 2, the canonical path P_{AB} between A and B exists in G' and has cost $cost(P_{AB}) \leq |AB|^p$ by part 5 of Theorem 1. Now the path P in G' from C to B consisting of the edge CA followed by the path P_{AB} has cost $cost(P) \leq |CA|^p + |AB|^p \leq (1 + 2^p \sin^p (\pi/k))|CB|^p$ by Lemma 2 ($\alpha = 2\pi/k$), as desired.

To show that the degree of G' is bounded by $k + 5$, note that at least one cone will be properly contained in every S_1 sector (since the angular sector determined by an S_1 sector is at least $(k - 2)\pi/2k \geq 2(2\pi/k)$ for $k \geq 12$), and at least ℓ cones will be properly contained in every $S_{\geq 2}$ sector with ℓ edges (since $\ell\pi/4 \geq (\ell + 1)2\pi/k$ for $k \geq 12$ and $\ell \geq 2$). Since there are precisely k cones around a point, no more than k outgoing edges from a point are selected by the algorithm in steps (1), (2), and (3). In step (4), an additional x outgoing edges are selected only if x incoming edges have not been selected. No incoming edge in step (4) could contribute to the selection of two outgoing edges because, otherwise, the two outgoing edges would be part of a maximal sparse sector, and hence would have been selected by step (1). Since the in-degree of G is bounded by 5, the theorem follows.

5   The Extended Algorithm

The algorithm presented in this section handles the cases when the integer $k \in \{8, 9, 10, 11\}$. We assume that the input to the algorithm is an oriented Gabriel graph G of in-degree bounded by 5, as described in the previous section. The main idea again is to ensure that all the outgoing edges in the S_1 and $S_{\geq 2}$ sectors around any point in G are selected among the (at most) k selected outgoing edges from that point. This ensures that, for any two points A and B in G, if the edge $AB \notin G$, then the edges on the canonical path between A and B in G will be selected. With the value of $k \geq 12$, this condition was easily satisfied since the cone size $(2\pi/k)$ was small enough to ensure that, given any fixed partitioning of the area around A into cones of size $2\pi/k$, each S_1 sectors contains at least one empty cone within it, and each $S_{\geq 2}$ sector with ℓ edges contains at least ℓ cones within it. This allowed us to pick all the outgoing edges in the S_1 and $S_{\geq 2}$ sectors around A, while ensuring that for any outgoing edge e from a point A, there is a selected outgoing edge e' from A with $\angle e, e' \leq 2\pi/k$, and without picking more than k outgoing edges from A.

When $k \in \{8, 9, 10, 11\}$, the cone size becomes larger and the above condition can no longer be satisfied. For instance, suppose that $k = 8$, and suppose that there are three $S_{\geq 2}$ sectors around a point A, each consisting of two outgoing edges making an angle of precisely $\pi/2$. Suppose further that each of the three angular sectors separating any two of the three $S_{\geq 2}$ sectors measures precisely $\pi/6$, and contains shorter outgoing edges than the $S_{\geq 2}$ edges. Then one readily sees that picking the six $S_{\geq 2}$ edges leaves us with only two edges to pick. This makes it impossible to pick outgoing edges from the three regions separating the $S_{\geq 2}$ sectors so that to guarantee that for every outgoing edge AB there is a selected outgoing edge AC with $|AC| \leq |AB|$ and such that $\angle BAC \leq 2\pi/k = \pi/4$. Observe here that, even though for any edge AB lying in a region between the $S_{\geq 2}$ sectors there is a selected edge that makes an angle of at most $\pi/4$ with AB, namely one of the boundary edges of the $S_{\geq 2}$ sectors, this fact does not help satisfying the desired condition because the boundary edge of the $S_{\geq 2}$ may be longer than AB.

To fix this problem, we first draw the following observation. The reason why we needed to satisfy the condition that, for every edge CB (the property now is applied to a point C) there is a selected edge CA such that $|CA| \leq |CB|$ and $\angle BCA \leq 2\pi/k$, is that we wanted to guarantee that $|CA|^p + |AB|^p \leq (1 + 2^p \sin^p (\pi/k))|CB|^p$ by Lemma 2, where $(1 + 2^p \sin^p (\pi/k))$ is the desired stretch factor. It turns out that even if the selected edge $|CA| > |CB|$, the inequality $|CA|^p + |AB|^p \leq (1 + 2^p \sin^p (\pi/k))|CB|^p$ is still satisfied as long as $\angle BCA$ is smaller than a small "approximation" angle α_{apx}^k that depends on k. Certainly this approximation angle will be smaller than $2\pi/k$. We will show next how a lower bound on the angle α_{apx}^k can be computed. With this in mind, we can now make use of the selected boundary edges of the S_1 and $S_{\geq 2}$ sectors to cover an additional angular sector of size α_{apx}^k each. We first compute a lower bound on α_{apx}^k for each value of $k \in \{8, 9, 10, 11\}$.

If $k = 8$ we set $\alpha_{apx}^k = \pi/12$; if $k = 9$ we set $\alpha_{apx}^k = \pi/14$; if $k = 10$ we set $\alpha_{apx}^k = \pi/17$; if $k = 11$ we set $\alpha_{apx}^k = \pi/22$. We have the following lemma.

Lemma 3. *For any $k \in \{8, 9, 10, 11\}$ let α_{apx}^k be as defined above. Let C, B, A be points in a Gabriel graph G such that CB and CA are edges in G, and such that $\angle BCA \leq \alpha_{apx}^k$. Then for any constant $p \in [2, 5]$, we have $|CA|^p + |AB|^p \leq (1 + 2^p \sin^p (\pi/k))|CB|^p$.*

Now we are ready to present the extended algorithm **Extended_Edge_Selection**. We assume that the input to the algorithm is an oriented Gabriel graph G of in-degree bounded by 5, as described in the previous section, and a parameter $k \in \{8, 9, 10, 11\}$. The algorithm will then compute a spanning subgraph G' of G whose edges are the edges of G selected by the algorithm. We first color the points in G with in-degree 0 black and all other points white. When an unprocessed point A is colored black it performs the following algorithm. Assume that $Out(A)$ is the set of all outgoing edges from A.

The Algorithm Extended_Edge_Selection

(1) **for** every maximal sparse sector around A **do** select the sector edges in $Out(A)$;

(2) **for** every unselected edge $AB \in Out(A)$ such that there exists a selected incoming consecutive edge CA with $\angle CAB \geq (k-2)\pi/2k$ and such that CA is not a sector edge in any maximal sparse sector around A **do** select AB;

(3) **for** every unselected edge $AM \in Out(A)$ such that there exists an unselected consecutive edge NA incoming to A with $\angle NAM \geq (k-4)\pi/2k$ **do** select AM;

(4) let $Unselected(A)$ be the set of all edges in $Out(A)$ that are still unselected; remove from $Unselected(A)$ every edge e such that the angle between e and a selected edge in $Out(A)$ is bounded by α_{apx}^k;

(5) let S be the sequence of remaining edges in $Unselected(A)$ in a clockwise (or anticlockwise) order;

(6) **while** $S \neq \emptyset$ **do**: place a cone of size $2\pi/k$ at the first edge in the sequence S, select the shortest edge in this cone, and remove all the edges in this cone from the sequence S;

(7) set the status of every unselected edge in $Out(A)$ to unselected and send a message to every outgoing neighbor B (i.e., AB is an outgoing edge) notifying it of whether the edge AB has been selected or not, and set the status of A to processed.

Upon receiving a message from a neighbor B, point A performs the following steps:

(1') update the status of the edge BA according to the message received from B;

(2') **if** for every incoming neighbor B (i.e., BA is an incoming edge) the status of the edge BA has been determined **then** color$(A) := $ black.

Lemma 4. *Let G be an oriented Gabriel graph, let A be a point in G, and let $k \in \{8, 9, 10, 11\}$ be a fixed constant. The following are true after the algorithm* **Extended_Edge_Selection** *finishes execution.*

(a) All the outgoing edges from A in the $S_{\geq 2}$ sectors around A are selected by the algorithm.

(b) All the outgoing edges from A in the S_1 sectors around A in which the incoming edge to A in the S_1 sector is not is not a sector edge in any maximal sparse sector around A, are selected by the algorithm.

(c) Every outgoing edge AM such that there exists an unselected incoming consecutive edge NA with $\angle NAM \geq (k-4)\pi/2k$ is selected by the algorithm.

(d) For every outgoing edge e from A in G, either there exists a selected outgoing edge e' from A not longer than e such that the angle $\angle e, e' \leq 2\pi/k$, or there exists a selected outgoing edge e' from A such that $\angle e, e' \leq \alpha_{apx}^k$.

(e) At most $k+5$ edges incident on A are selected by the algorithm.

Let G' be the subgraph of G whose point set is the point set of G, and whose edges are the edges selected by the above algorithm when applied to G.

Theorem 4. *The subgraph G' is a power spanner of G of maximum degree $k+5$ and a stretch factor $\rho \leq 1 + 2^p \sin^p (\pi/k)$.*

Proof. By part (e) of Lemma 4, every point in G' has degree bounded by $k + 5$. This shows that G' has a maximum degree $k + 5$.

To prove the upper bound on the stretch factor, it suffices to show by Lemma 1 that every edge in G has a stretch factor bounded by $1 + 2^p \sin^p (\pi/k)$ in G'. Let CB be an edge in G. If CB is an edge of G' then we are done. Suppose now that CB is not an edge in G'. Then CB was not selected by the algorithm. Without loss of generality, assume that the edge CB was directed from C to B in the oriented graph G (otherwise, exchange C and B). By part (d) of Lemma 4 applied to point C, either an edge CA outgoing from C is selected with $|CA| \leq |CB|$ and such that the angle $\angle BCA \leq 2\pi/k$, or an edge CA outgoing from C is selected such that $\angle BCA \leq \alpha_{apx}^k$.

Case I: *An edge CA outgoing from C is selected with $|CA| \leq |CB|$ and such that the angle $\angle BCA \leq 2\pi/k$.* Since $|CA| \leq |CB|$ and $\angle BCA \leq 2\pi/k$, and since the algorithm picks all the outgoing edges in the S_1 and $S_{>2}$ sectors around every point of G, by a similar proof to that of Theorem 2, the canonical path P_{AB} between A and B exists in G' and has cost $cost(P_{AB}) \leq |AB|^p$ by part 5 of Theorem 1. Now the path P in G' from C to B consisting of the edge CA followed by the path P_{AB} has cost $cost(P) \leq |CA|^p + |AB|^p \leq (1 + 2^p \sin^p (\pi/k))|CB|^p$ by Lemma 2 ($\alpha = 2\pi/k$).

Case II: *An edge CA outgoing from C is selected such that $\angle BCA \leq \alpha_{apx}^k$.* Since CA is a Gabriel edge in G the angle $\angle CBA$ is acute. Combining this fact together with the fact that $\angle BCA \leq \alpha_{apx}^k$, it can be easily verified using the values for α_{apx}^k for $k \in \{8, 9, 10, 11\}$, that the angle $\angle CAB \geq (k-2)\pi/2k$. Similarly, the angle $\angle CBA \geq \pi/2 - \alpha_{apx}^k \geq (k-4)\pi/2k$. With this fact in mind, it follows by a similar proof to that of Theorem 2, that the canonical path P_{AB} exists in G', and has cost $cost(P_{AB}) \leq |AB|^p$ by part 5 of Theorem 1. Now the path P in G' from C to B consisting of the edge CA followed by the path P_{AB} has cost $cost(P) \leq |CA|^p + |AB|^p$. By Lemma 3, $|CA|^p + |AB|^p \leq (1 + 2^p \sin^p (\pi/k))|CB|^p$, and $cost(P) \leq |AC|^p + |AB|^p \leq (1 + 2^p \sin^p (\pi/k))|CB|^p$.

6 Concluding Remarks

Song et al. [8] presented two localized distributed algorithms for computing a bounded-degree planar power spanner of a unit disk graph. The first one computes a planar spanner with bounded degree $k + 5$ and a stretch factor bounded by $1/(1 - 2^p \sin^p (\pi/k))$, for any integer parameter $k \geq 7$. The second one computes a planar spanner with bounded degree k and a stretch factor bounded by $(\sqrt{2})^p/(1 - (2\sqrt{2} \sin (\pi/k))^p)$, for any integer parameter $k \geq 9$. The algorithm presented in the current paper significantly improves the algorithms in [8] for all values of $k \geq 8$ and $p \in [2, 5]$, as illustrated in Table 1.

Table 1. Comparison between the algorithm in this paper and those in [8] for $p = 2$

$k =$	8		9		10		11		12	
	Δ	ρ	Δ	ρ	Δ	ρ	Δ	ρ	Δ	ρ
The algorithm in this paper	13	1.585	14	1.467	15	1.381	16	1.317	17	1.267
The first algorithm in [8]	13	2.414	14	1.879	15	1.618	16	1.465	17	1.366
The second algorithm in [8]	13	3.691	14	3.311	15	3.057	16	2.875	17	2.740

While our algorithm improves upon previous algorithms, it is not known what the optimal stretch factor is for a given degree bound k. In particular, the $+5$ term in the degree bound resulting from the orientation of the graph seems artificial, and it is possible that this term can be avoided using different techniques that do not take advantage of the ordering on the vertices introduced by this orientation. It remains an open problem whether there exists a distributed localized algorithm improving the stretch factor/degree bound combination given in this paper.

References

1. Prosenjit Bose and Pat Morin. Online routing in triangulations. *SIAM J. Comput.*, 33(4):937–951, 2004.
2. K. Ruben Gabriel and Robert R. Sokal. A new statistical approach to geographic variation analysis. *Systematic Zoology*, 18(3):259–278, 1969.
3. M. Grünewald, T. Lukovszki, C. Schindelhauer, and K. Volbert. Distributed maintenance of resource efficient wireless network topologies. volume 2400 of *Lecture Notes in Computer Science*, pages 935–946, 2001.
4. X.-Y. Li, P.-J. Wan, Y. Wang, and O. Frieder. Sparse power efficient topology in wireless networks. *Journal of Parallel and Distributed Computing*. To appear.
5. Xiang-Yang Li, Peng-Jun Wan, and Yu Wang. Power efficient and sparse spanner for wireless ad hoc networks. International Conference on Computer Communications and Networks, pages 564 – 567, 2001.
6. Rajmohan Rajaraman. Topology control and routing in ad hoc networks: a survey. *SIGACT News*, 33(2):60–73, 2002.
7. Ram Ramanathan and Regina Hain. Topology control of multihop wireless networks using transmit power adjustment. In *INFOCOM 2000*, pages 404–413, 2000.
8. Wen-Zhan Song, Xiang-Yang Li Yu Wang, and Ophir Frieder. Localized algorithms for energy efficient topology in wireless ad hoc networks. *Mobile Networks and Applications*, 10(6):911–923, 2005.
9. Yu Wang and Xiang-Yang Li. Localized construction of bounded degree and planar spanner for wireless ad hoc networks. Joint Workshop on Foundations of Mobile Computing, pages 59 – 68, 2003.
10. Yu Wang, Xinag-Yang Li, and Ophir Frieder. Distributed spnanner with bounded degree for wireless networks. *International Journal of Foundations of Computer Science*, 14(2):183–200, 2003.
11. Andrew Chi-Chih Yao. On constructing minimum spanning trees in k-dimensional spaces and related problems. *SIAM Journal on Computing*, 11(4):721–736, 1982.

Wireless Communication in Random Geometric Topologies

Luděk Kučera[1] and Štěpán Kučera[2]

[1] Charles University, Prague, Czech Republic
[2] Kyoto University, Kyoto, Japan

Abstract. The present paper deals with communication in random geometric topologies; in particular, we model modern wireless ad hoc networks by random geometric topologies. The paper has two goals: the
first is to implement the network power control mechanism extended
by several wireless engineering features and to use the model to assess
communicational properties of two typical random geometric topologies
known from the literature in real-life conditions. The second goal is to
suggest a modification of the "LowDegree" algorithm (one of the two
studied topologies), which preserves excellent power requirements of the
model, but matches the performance of the standard "UnitDisk" algorithm in terms of higher total network throughput.

1 Introduction

Recently, mobile ad hoc networks (also known as distributed radio networks)
have received much attention as they are regarded as the next frontier systems
for ubiquitous mobile multimedia communications. However, their operation also
imposes some major technical challenges, in particular the problem of distributed
control and management of radio resources, since an ad hoc network cannot be
controlled by a centralized controller as in the case of nowadays cellular networks
and moreover ad hoc network resources such as energy or bandwidth are often
scarce and effective system management is essential to promote the quality and
efficiency of system services [5].

In this context, we note that random geometric graphs (see e.g. [10]) represent
natural means for modeling modern radio network with random topologies and
are intensively studied due to their wide application possibilities in the communication theory by both mathematicians and engineers as they seem to be
promising concept.

Wireless ad hoc networks are characterized by several aspects, which are
unique in domain of communication engineering. Unlike wired networks with
invariable transmission media such as e.g. computer/telephony or optical networks, wireless network designers face specific design challenges and have to
among others mitigate interference, maximize achievable throughput capacity
and maintain to some extent link quality of service with respect to node mobility and channel impairments, while minimizing power consumption and increasing the battery life [6]. For these reasons, many solution concepts from wired
networks cannot be successfully applied and new approaches are necessary [12].

S. Nikoletseas and J.D.P. Rolim (Eds.): ALGOSENSORS 2006, LNCS 4240, pp. 107–118, 2006.
© Springer-Verlag Berlin Heidelberg 2006

It is clear from the above mentioned that when looking for such transmit powers, which would assure a connected network, it also is necessary to pay attention to the existence of a characteristical trade-off of wireless engineering. Said trade-off basically consists of finding a balance between (i) increasing transmit power of network nodes in order to assure connectivity and reliable and fast transmissions, and (ii) decreasing the undesirable interferences in a shared wireless channel.

It is a well-known fact that higher transmit power imply better propagation of electromagnetic waves to more distant places and leads - for a given interference and/or noise level - to an increase of the communication channel capacity [11]. On the other hand, lower transmit powers often increase the network efficiency in terms of transmitted bits per Joule of used energy and network/node lifetime as network nodes have typically only limited power supplies. More importantly however, lower transmit power also imply lower interference in the case that more transmitters are active simultaneously. This can in turn increase network data throughput and higher spatial reuse of the network topology.

Therefore, the principal purpose of the connectivity problem related power control schemes in wireless networks cannot be only obtaining good network connectivity at low costs, but it must also allow nodes to setup and maintain wireless links without causing unnecessary interference to others while satisfying constraints on quality of service while adapting to network node movements, fluctuating interference, channel impairments, and so on [2]. Power control also represents an efficient way to improve the spatial reuse by allowing multiple pairs to communicate simultaneously [3].

Moreover, energetic conditions in a wireless networks have also a significant impact on the transport layer (of the OSI model), particularly on routing methods, which take into account transmit powers, remaining battery energies or other energy related parameters - e.g. minimum power routing and/or network lifetime maximizing routing [9]. Furthermore, power control has impact to data link layers of the OSI system model, as it can be used for implementing general admission control or particular realization of cellular network operations such as channel selection and switching and hand-off control [2].

Nevertheless, we note that there is a substantial lack of detailed studies, proposing well performing power control schemes, which would not only achieve a required kind of network connectivity, but would be evaluated also from the point of view of the resulting interference for different topologies under real (or quasi-real) wireless network conditions. While the LowDegree algorithm proposed in [8] is excellent in achieving connectivity with extremely low (asymptotically optimal, cf. [4]) power, it behaves poorly because of substantial congestion and stretch factor. We therefore propose a modified power control algorithm that matches more robust models (e.g., Unit Disk) in performance while being substantially more power efficient. A wireless engineering study brings also more insight into properties of standard and well-known topologies that have not yet been studied in detail from the point of view of spatial reuse and interference.

2   Connectivity Optimizing Power Control Schemes

We study models of communication among users that are represented by random points of a square are in the two-dimensional plane. Given a finite set $X \subset [0, S]^2$ for some real $S > 0$, a power assignment is a function $p : X \to [0, \infty)$. A power assignment defines an oriented graph on the set X in the following way: given $u, v \in X$, (u, v) is an oriented edge iff $\ell^2(u, v) \leq p(u)$, where $\ell(u, v)$ is the Euclidean distance of points u and v. Quite often it is convenient to investigate symmetric (unoriented) graphs, where edges are bi-directional, because this makes it possible to use full duplex communication in any link. In such a case (u, v) is an edge iff $\ell^2(u, v) \leq \min(p(u), p(v))$.

Given a node $u \in X$, $p(u)$ means the power of the transmitter associated with the node u. Assuming homogenous (constant) noise and equal quality of receivers, a node u can transmit into a distance that is proportional to the square root of the power.

If we need to send a packet to a distant node that cannot receive a signal of the transmitter directly, we would send it along a path that consists of several hops - edges of the graph over the set X. It is therefore necessary to assume that the graph over a set of points is connected.

Good connectivity can be obtain when using high transmit powers. However, mobile networks are often power sensitive (e.g., if nodes of a network are sensors, then we have to assume very limited power resources). Moreover, which is much more important, if several nodes are transmitting simultaneously, high transmit powers cause high amount of interference that can make efficient communication impossible.

As a measure of power requirements we will use the quantity $\sum_{u \in X} p(u)$, the sum of all node powers.

Two extreme topologies are investigated in the present paper. The first one, usually called the unit disk model (or, more precisely, the constant diameter disk model) assumes that the power assignment function is constant. This means that disks around nodes that determine possibility of direct transmission have the same radius for any node. If the node powers are equal to a constant p, we will denote this graph by $U(X, p)$.

Thus, given a set $X \subset [0, 1]^2$, the optimal connected unit disk graph over X is a graph $U(X, p_{conn})$, where p_{conn} is the smallest positive real number p such that $U(X, p)$ is connected. We will use the name *UnitDisk* to identify this model.

The unit disk model is very popular because of its simplicity, and there is a vast literature that deals with the model, see [10] and the references presented there. Unfortunately, it has been proved that the expected sum of node powers in a connected unit disk graph $U(X, p_{conn})$ is almost surely $\Theta(S^2 \log n)$, where S is the size of the bounding square, and n is the number of nodes, while $\Theta(S^2)$ is the lower bound of the sum of powers expectation [4]. Moreover, there are power assignment algorithms that can almost surely match the lower bound up to a constant multiplicative factor. This means that a unit disk model is almost surely an inefficient way to interconnect a random set of points in a square area.

Another frequent model is k-nearest neighbor graph ($k - NNG$) model [14]. In this model, the power of a node u is set to the square of the distance from u to its k-nearest neighbor in the set X. In other words, a node u is connected to its k nearest neighbors. However, the model has properties similar to the unit disk and hence it is not considered in the shortened version of the paper.

On the other hand, we will investigate a topology that was suggested in [8] and will be called *LowDegree*. The graph construction is motivated by the $k - NNG$ models and proceed in rounds for $k = 0, 1, 2, 3, \ldots$. Nodes that are active in the k-th round increase, if necessary, their power to reach k nearest neighbors and, if necessary, the k-th nearest neighbor of u (i.e., a new neighbor of u) must increase its power to be able to call back to u. As a result of the process, a graph, which originally (in the 0-th round) consists just isolated nodes, is composed of connected components that collate together. A node is active until the size of its component involves more than one half of all nodes, then it stops increasing transmit power. The model is symmetric, the call back clause guarantees that an edge to the k-th nearest neighbor is an edge even in the symmetric model. For details, see [8].

It has been shown in [8] that the expected power requirements of this model are $\Theta(S^2)$ (i.e., the bound does not depend on the number of nodes), which means that the model is asymptotically optimal up to a constant multiplicative factor, see [4].

Fig. 1 shows an example of a unit disk graph (left) and LowDegree graph (right) on the same set of 250 nodes.

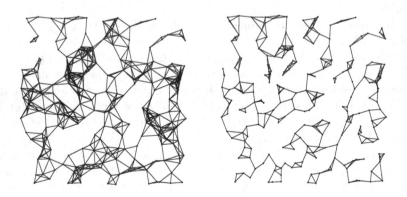

Fig. 1. UnitDisk graph and LowDegree graph with 250 nodes

It is clear that the LowDegree topology is much more power efficient as a way of interconnecting a set of points in a square area than the UnitDisk topology. On the other hand, the LowDegree model shows that connectivity is a necessary, but not sufficient condition for successful communication in a wireless network.

As it can be seen from Fig. 1, a path connecting different nodes of a LowDegree graph is usually very long, and the stretch factor (the ratio of the distance of a pair of nodes in the network and their Euclidean distance) is very high for many node pairs. Consequently, both the number of nodes that have to be active to deliver a single packet to its destination is quite high, and the communication is also slowed down by high congestion in a network.

One of the main contributions of the present paper is a modification of the LowDegree topology that preserves low power consumption of this model, but improves highly connectivity properties like shorter paths and lower average stretch factor, higher number of disjoint path connecting nodes (higher degree of connectivity) and decreased congestion.

The modification consists of insertion of *bridges*: a graph with k bridges results from the following procedure:

---

build a LowDegree graph over a given set of points;
repeat the following k times
       find a pair u, v of two different nodes that maximizes the stretch;
       increase the power of u and v so that
            both (u, v) and (v, u) become edges of the graph.

Construction of bridges

Note that the stretch of node pairs is recomputed after any bridge insertion.

Fig. 2 shows an example of a LowDegree network on 250 nodes (left) and the same graph extended by 30 bridges (right).

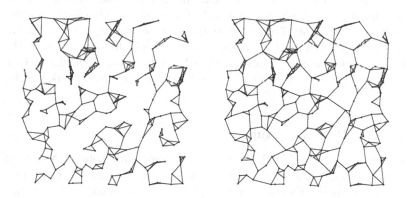

Fig. 2. LowDegree graph with 250 nodes and the same graph with 30 bridges

It is clear that a graph with moderate number of bridges remains quite sparse (which means that the sum of transmit powers is low), but connectivity of the network with bridges is much better.

As defined above, a bridge insertion is an idealized operation, but some more realistic modifications of the simple bridge insertion have been found to yield essentially the same results (not covered by the paper).

3 System Model

In this section, we describe the system model of a wireless ad hoc network, which we used for an overall performance evaluation of a real-life ad hoc network in terms of throughput and required energetic demands. We used the proposed LowDegree algorithm and for comparison purpose also the UnitDisk algorithm as the power control scheme on the physical layer of the OSI model. The data link layer was implemented using the CSMA/CA technique and on the network layer, we used simple minimum power routing. Generally speaking, we try to keep our simulation close to the popular IEEE 802.11a wireless standard in ad hoc mode.

3.1 Channel Orthogonalization

A general approach to modeling of wireless multihop networks consists in abstracting a multihop wireless network into a set of several mutually orthogonal (non-interacting) channels, in which several simultaneously active links can be accommodated. Links correspond to concurrent single-hop transmissions. The orthogonality among different channels can be achieved by many proposed division schemes, using or combining techniques of time-division multiple access (TDMA) or frequency-division multiple access (FDMA).

It is therefore reasonable to analyze the interaction of multiple links in any one of such isolated communication channels due to its interference orthogonality with other bands. Within one orthogonal channel, different links can use code-division multiple access technique (CDMA) for separating ongoing transmissions, which however does not eliminate the necessity of power control due to the near/far effect and others [13]. In this work, we will follow the same channel isolation paradigm and study only a reduced problem of a single channel CDMA network [7] or a single FDMA channel network with concurrent link interference [2].

3.2 General Network Setup

We consider a square network of S by S meters with totally N transceiver nodes (able to transmit or receive data), comprising m randomly chosen source-destination node pairs. Each source node wants to send a constant amount D of data to its corresponding destination node.

If a source node cannot reach its destination node in one hop - for example because its allocated transmit power would not be sufficient to guarantee a reliable connection, the source node data would be forwarded by intermediate relaying node(s), repeatedly forwarding the source data via a multihop connection until its corresponding destination. Each network node can either transmit *or* receive data, but not both at the same time.

We assume a multi-user communication in orthogonal CDMA or FDMA channels with concurrent link interference [2]. We denote by h_{ij} the power loss from the transmitter i to the receiver j. This comprises path loss, multipath fading, shadowing and other radio propagation effects, as well as spreading gain/processing gain of CDMA transmission. To keep things simple, we will assume all channel gains h_{ij} deterministic and time invariant as a result of time averaging and no mobility of links respectively. A more complicated case of stochastic analysis of time-varying processes is not relevant for the scope of this contribution. In more detail, we assume the path loss to be equal to $1/d_{ij}^{\alpha}$, where d_{ij} is the distance between transmitter i and receiver j and α is the path loss exponent $\alpha \in [2; 4]$.

3.3   Traffic Model

Every user i is assigned some transmit power P_i^{TX}. We use our LowDegree algorithm, optionally with a predefined number of bridges, to calculate the necessary transmit power values to achieve a low degree connectivity and compare its achieved network performance with a model, in which power control is realized by the UnitDisk algorithm.

In either cases, two nodes i and j are "connected" if the power level $P_j^{RX} = h_{ij}P_i^{TX}$, received at the receiver j from the transmitter i through the communication channel, is equal to some constant P_{req}^{RX}, which denotes a required received power level. In the case that a source node cannot establish a connection to its destination according to this definition, its data would be forwarded through a minimum transmit power path, i.e. through forwarding nodes, whose sum of allocated transmit powers would be the smallest possible for the given network topology and allocated transmit powers. This routing metrics is beneficial from the energetic point of view, yet can lead to increased congestion. For other transmit power related metrics, which solve the increased congestion problem by sacrificing some of the energetic savings, see e.g. [9]. We note however that the particular choice of the routing metrics is not relevant to the main idea of this contribution.

Each source node sends its D packet to its destination only once. After all sources send their data to their destinations through corresponding routing paths, the simulation is over.

3.4   Admission Model

On any routing path, if a node is transmitting or receiving data, any other node, which would like to send data to this busy node, must wait until this node gets free and terminates its transmission or reception activity. The availability of the "next hop" node is not the only criterion of allowing a node to transmit it data.

A node j will start to transmit its data only if the received interference at its location, given by the sum of received power from all active transmitters i in the network $\sum_{allTXsi} h_{ij}P_i^{TX}$, is smaller than some maximum interference level I_{max}. If however $\sum_{allTXsi} h_{ij}P_i^{TX} > I_{max}$, then this node must wait until the received interference level decreases below I_{max}. This can happen if some close transmitter finishes its transmission session.

Such an interference aware admission scheme implements in fact so called "carrier sensing multiple access with collision avoidance" random access technique (CSMA/CA), which is a random access technique, well-known in the wireless engineering as the currently best performing technique for the considered network model. Roughly speaking, each potentially transmitting node cannot start to transmit if the interference in the shared channel is too high, since a sudden transmission would most probably lead to a packet collision and possible packet lost and undesirable retransmissions.

A packet of size D, sent by a transmitter i, is successfully received by a transmitter j, when

$$\int_{T_{begin}}^{T_{end}} C\left(\gamma(t)\right) dt = D \tag{1}$$

where T_{begin} and T_{end} are times of the beginning and end of the transmission respectively, whereby the capacity of a communication channel with Rayleigh fading $C\left(\gamma(t)\right)$ [1] is given by

$$C\left(\gamma(t)\right) = Blog_2(e)e^{1/\gamma(t)}\left(-E + ln(\gamma(t)) - \sum_{k=1}^{+\infty}\frac{(-1/\gamma(t))^k}{k \cdot k!}\right) \tag{2}$$

with B being the frequency bandwidth, $E = 0.5772$, $\gamma(t)$ being the signal to interference and noise ratio, defined for transmitter i, receiver j and all surrounding active transmitters $-i$ as $\gamma(t) = \frac{h_{ij}P_i^{TX}}{\sum_{allTXs-i}h_{-i,j}P_{-i}^{TX}}$ and P_e being bit error rate (given by a particular coding and modulation technique with respect to a given γ). Note that $\gamma(t)$ is a function of time as the set of active transmitters is changing in time. On the other hand, a D packet is lost if $\gamma(t) < \gamma_{min}$ for some time period T_{min}.

The product of the Rayleigh channel capacity C and the bit error rate P_e gives roughly speaking the theoretical upper bound for the data rate in a mobile wireless data link, which is a subject to fast fading - a typical propagation effect of electromagnetic waves in a mobile wireless environment. Such a model represents the state-of-the-art model of a mobile ad hoc wireless network, using the latest results of wireless engineering.

For comparison, results of simulations based on the classical Shannon channel capacity are included as well. As it will be clear, the two model differ only insignificantly (as expected, a Rayleigh channel being always a bit slower than the Shannon one).

4   Simulation Setting

We have generated random topologies with 256 nodes (more sizes in the full version of the paper), whereby for each topology, we have randomly chosen 1000 source-destination pairs. All sources were simultaneously trying to send one message of size D to their respective destination following above described admission and routing rules, while using powers, assigned by either the LowDegree algorithm or the UnitDisk algorithm.

We compared the LowDegree without and with bridges and the UnitDisk power/ connectivity control schemes by means of comparing the overall performance of the resulting networking system in given three topologies.

In our simulations, the following quantities are evaluated: (i) total time the network needs to deliver messages of the size D along all its 1000 routing paths, (ii) average length of said routing paths (i.e., the number of hops), (iii) average number of active transmitters in the network and (iv) the total power consumed by the network when all nodes were active at the same time. We observe the evolution of these quantities with respect to increasing the number of bridges from 0 to 50 in the LowDegree power assignment scheme and compare them with levels achieved by the UnitDisk network.

In the plot showing the time necessary to deliver a given number of messages in a network two curves are shown. The upper one (longer times) corresponds to the Rayleigh channel from Eq. 2; for comparison we also show results for Shannon capacity of an additive white Gaussian noise channel [11].

5   Simulation Results

As we have already noted, the LowDegree algorithm is assigning much smaller transmit powers than the UnitDisk one, resulting is much lower total power usage and therefore better performance on the physical layer. Nevertheless from the point of view of the *overall* network performance, the LowDegree network is usually several times slower than the UnitDisk one (see Fig. 3, the number of bridges equal to 0). This is due to the fact that on the routing level, the LowDegree algorithm assigns low powers, leading to a very sparse connectivity in comparison with the UnitDisk algorithm, which results in turn in relatively long routing paths, causing eventually undesirable delivery delay and thus makes the delivery time of the LowDegree network considerably longer than the one of the UnitDisk network.

However, after insertion of even a very small number of bridges both networks become rapidly approximately equally fast, while the LowDegree network with bridges has only slightly higher power requirements compared to the case with no bridges, i.e., the LowDegree network with bridges has about the same delivery time as the UnitDisk network, yet it is using substantially lower total power to achieve this goal (compare with Fig. 4).

The above mentioned fact that the LowDegree power assignment achieves network connectivity in an energetically efficient way (efficient physical layer), but also results in longer routing paths (inefficiency on the network layer), is clearly visible from Fig. 5 - the average path length is longer in the LowDegree network with no bridge than in the UnitDisk. However we can see that the bridges insertion rapidly equalizes this difference between the two algorithms. The rate of the response is analogical to the decrease of the delivery time from

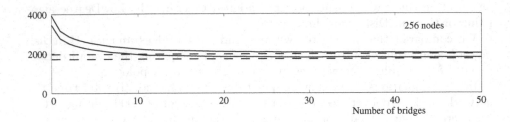

Fig. 3. Average total time needed to deliver one D message (network topologies with 64, 256 and 1024 nodes)

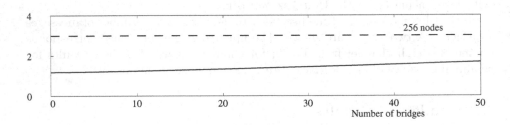

Fig. 4. Total power consumed in the network during a simulation run (network topologies with 64, 256 and 1024 nodes)

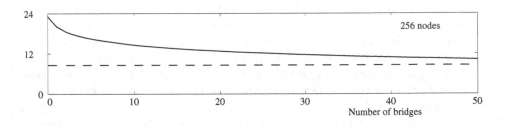

Fig. 5. Average path length in the network (network topologies with 64, 256 and 1024 nodes)

Fig. 3 and the cross-point between graph lines is located in time only on an slightly higher position.

In accordance with Fig. 6, the average number of active transmitters, which were admitted to the active state by the CSMA/CA mechanism based on an evaluation of local interference level, is higher for the LowDegree network than for the UnitDisk one and moreover does not depend too much on the number of bridges. It decreases only very slowly with the number of bridges. We can clearly see that the LowDegree algorithm results in lower interference that the UnitDisk one, which is a key property in wireless networks design. We also can see from

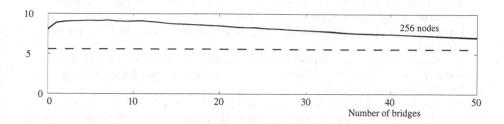

Fig. 6. Average number of active nodes in the network under CSMA/CA admission scheme (network topologies with 64, 256 and 1024 nodes)

our simulations that on the cross-layer performance level, the higher average path length and higher number of active transmitters roughly compensate each other.

Lastly, we briefly note that the capacity evaluation for the Rayleigh channel from Eq. 2 gives expectedly proportionally lower values in both LowDegree and UnitDisk networks if compared with Shannon capacity of an additive white Gaussian noise channel [11]. This fact justifies that our simulations hold for both mobile wireless environment (approximation by Rayleigh channels) and standard additive white Gaussian noise channel.

Acknowledgement

The research was supported by the European Commision - Fet Open Project DELIS IST-001907. The first author was also supported by Czech Ministry of Education, Youth and Sports grant No. MSM0021620838.

References

1. Mohamed-Slim Alouini and Andrea J. Goldsmith. Capacity of rayleigh fading channels under different adaptive transmission and diversity-combining techniques. *IEEE Trans. Veh. Technol.*, 48(4):1165–1181, July 1999.
2. Nick Bambos. Toward power-sensitive network architectures in wireless communications: Concepts, issues, and design aspects. *IEEE Personal Commun. Mag.*, 5(3):50–59, June 1998.
3. Q. Chen and Z. Niu. A game-theoretical power and rate control for wireless ad hoc networks with step-up price. *IEICE Transactions on Communications*, E88-B:3515–3523, September 2005.
4. A. Clementi, P. Penna, and R. Silvestri. On the power assignment problem in radio networks. *Mobile Networks and Applications*, (9):125–140, 2004.
5. Andrea Goldsmith. *Wireless Communications*. Cambridge University Press, 2005.
6. Andrea J. Goldsmith and Stephen B. Wicker. Design challenges for energy-constrained ad hoc wireless networks. *IEEE Trans. Wireless Commun.*, 9(4):8–27, August 2002.

7. T. C. Hou and V. O. K. Li. Transmission range control in multihop packet radio networks. *IEEE Trans. Commun.*, 34(1):38–44, January 1986.
8. Ludek Kucera. Low degree connectivity in ad-hoc networks. In *ESA*, pages 203–214, 2005.
9. A. Michail and A. Ephremides. Energy-efficient routing for connection-oriented traffic in wireless ad hoc networks. *Mobile Networks and Applications 8*, pages 517–533, August 2003.
10. M. Penrose. *Random Geometric Graphs*. Oxford University Press, Oxford, 2003.
11. John G. Proakis. *Digital Communications, 4th Edition*. Irwin/McGraw-Hill, 2001.
12. V. Raghunathan and P. R. Kumar. A counterexample in congestion control of wireless networks. In *Proceedings of the 8-th ACM/IEEE International Symposium on Modeling, Analysis and Simulation of Wireless and Mobile Systems*, pages 290–297, Montreal, Canada, Oct. 10-13 2005.
13. Sergio Verdú. *Multiuser Detection*. Cambridge University Press, New York, 1998.
14. F. Xue and P. R. Kumar. The number of neighbors needed for connectivity of wireless networks. *Wireless Networks*, (10):169–181, 2004.

Localization Algorithm
for Wireless Ad-Hoc Sensor Networks
with Traffic Overhead Minimization
by Emission Inhibition

Pierre Leone, Luminita Moraru*, Olivier Powell**, and Jose Rolim

Department of Informatics,
University of Geneva, Switzerland
{leone, moraru, powell, rolim}@cui.unige.ch

Abstract. Widely used positioning systems like GPS are not a valid solution in large networks with small size, low cost sensors, due both to their size and their cost. Thus, new solutions for localization awareness are emerging, commonly based on the existence of a few references spread into the network.

We propose a localization algorithm to reduce the number of transmitting nodes. The algorithm relies on self selecting nodes for location information disclosure. Each node makes a decision based on its proximity to the nodes in the area covered only by two of the references used for its own localization. We analyze different aspects of the location awareness propagation problem: communication overhead, redundant transmissions, network coverage.

1 Introduction

The knowledge of the geographical position is necessary in a variety of sensor networks applications and communication protocols. In a sensor network, each entity uses its sensing capabilities to collect information about the surrounding area and to send it periodically or event driven to a base station, in a hop by hop manner. Routing protocols either use location as criteria for building hierarchical architectures for data dissemination [1] or to identify the area to be monitored for a specific event [2].

Although accurate localization techniques, like GPS devices are currently available, their usage for networks with a large number of sensors is difficult due to the cost and the size of the necessary equipment. The alternative is to use a small number of GPS enabled nodes, considered as references by the rest of the nodes of the network to localize themselves.

* Research partially founded by the Swiss SER Contract No. C05.0030 and FP6-015964 AEOLUS.
** Research partially supported by the Swiss National Science Foundation grant no. 20021-1040107.

S. Nikoletseas and J.D.P. Rolim (Eds.): ALGOSENSORS 2006, LNCS 4240, pp. 119–129, 2006.

Current algorithms consider the aspect of accuracy and its impact on localization aware applications. One of the most important issues in sensor networks is the network lifetime. Since the main energy consumption source is represented by the communication mechanism [3], [4], [5], one of the main concern of any algorithm that require network traffic is to minimize the number of messages sent into the network.

We propose a method of minimizing the number of nodes that disclose their position. We consider an algorithm for self selection of emitters. During the localization process, each of the new localized node will emit only if it provides the best coverage for an area covered by any pair of its references. Our algorithm eliminates the message overhead of a set of nodes closely situated by selecting the one situated at the best position from the coverage point of view.

Our main results are to show through simualations that our algorithm succeeds in inhibiting all the transmissions which are not necessary for the localisation process to succeed, thus making it (in a weak sense) optimal, and to find a robust waiting time function (used in the inhibition process) which guarantees success of the localisation process in a short running time.

The paper is organized as follows. We begin with a presentation of the current work in localization of sensor networks in section II. Section III describes the details of the proposed algorithm. In section IV we validate our algorithm by the means of experimental results. Finally we present our conclusions in section V.

2   State of the Art

The localization process starts up with a small number of known position nodes, called anchors or beamers. They advertise their position in the network and the information is used by the other nodes to compute their own location. Based on the geographical position of the anchors and the estimated distance to them, a node can compute its own location, in a process called atomic multilateration.

Range based algorithms use different physical measurements for estimation of the distance to known location entities. In range based protocols [6],[7], [8] position awareness is propagated through the network during an iterative process. Each node, after receiving at least three known references, computes its location and became a source at the next step. The drawback of this algorithm is the message overhead and the increase probability of collision generated by simultaneous located nodes.

Range free algorithms approach this problem by using different methods to compute location without measuring the distance to the anchor. They use metrics, like for example the hop count to an anchor to estimate the coordinates, making the message overhead even higher. In [9], each anchor broadcasts a message into the network, containing its location. Each node waits for all the beamers and computes its location based on the centroid method that considers all known locations. The traffic overhead depends on the number of sources. DV-hop [10] keeps track of the number of hops to the reference and computes location based on the average distance between one hop neighbors. For each anchor, a message with the coordinates is broadcasted into the network. After the anchors collect information related to the coordinates and hop count of the entire set, they compute and broadcast into the network the average range. Compared with the centroid algorithm, DV-hop will include additional traffic.

3   Algorithm Description

The condition to initialise the algorithm is that 3 anchors nodes are close enough from each other to have common neighbours. The anchors start by broadcasting their position. A node is able to localize itself if it receives at least 3 location advertising messages. In an iterative process, like the greedy algorithm, every newly localized node broadcasts its position. Our aproach is to inhibit the redundant emissions.

Consider the case of three references (localised nodes), each of them in the range of the other two, as shown in Fig. 1(a). The area covered by a set of references is divided in three regions. The *3Covered* area contains the nodes which are in the transmission range of the three references: they will become location aware. The *2Covered* area contains nodes in the transmission range of each pair of references. The *1Covered* area contains nodes in the transmission range of only one reference. The neighborhood of the nodes and the division in several regions is presented in Fig. 1(b).

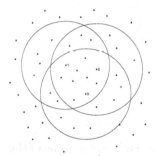 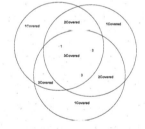 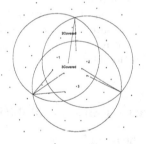

(a) The neighborhood of the initial anchors

(b) The regions of the covered area

(c) The evaluation of the potential references

Fig. 1.

Each node in the network listens for three location advertising messages. After it receives three messages, it is able to compute its geographical position (by using the *calculate_coordinates()* function).

In order to propagate the location awareness, each *2Covered* area needs a new emitter. The nodes in the *3Covered* area are good candidates for being this new emitter. Our idea is that a node should send its coordinates only if it is the closest node to one of the *2Covered* area. The distance evaluation is made by the *calculate_distances()* function. Given a newly localized node, we define its *critical point* as the closest point among the intersection points of the tree pairs of coverage circles, not adjacent to the *3Covered* area. The evaluation of the distances to the *critical point* is shown in Fig. 1(c).

A newly localised node sets a timer before deciding if it will broadcast or not, as discused in 3.1. If a node receives a message from a neighbour closer to its critical point before the timer expires, then its transmission is canceled and the timer is stopped and no message is sent, c.f. 3.2.

The algorithm is presented below:

Algorithm 1. Emitting decision algorithm

---
wait for 3 anchors
calculate_coordinates()
calculate_distances()
set_timer(distances)
while time_is_active **do**
  if new_message() **then**
    check_proximity()
    if closer **then**
      drop_timer()
      exit()
    end if
  end if
  if end_timer() **then**
    broadcast_coordinates()
  end if
end while

3.1   Emitters Inhibition

The main feature of the algorithm is the prevention of the emission of some of the nodes, while still completing the localization process. The purpose of a 3Covered area node emission is to provide a third source to the nodes from a 2Covered area. The inhibition takes place when a better transmission is heard, by dropping the timer and the broadcast. If a node hears a better emitter before its timer expires, it will decide not to emit. The *check_proximity()* function defines if an emitter is better or not.

A node will consider its distance to the critical point in order to take an inhibition decision. The emission inhibition with selection verifies if the emitter is situated closer to the same critical point as the current node, if this is so, it considers the new emitter to be better. An alternative is the emission inhibition without selection, where a node drops the counter if it hears a message from any of its neighbours, without previous verification of the proximity.

3.2   Time Selection Formula

A first approach is to randomly initialize the timer. It reduces the complexity of the algorithm, but better performances can be achived. Indeed, a better criteria is to base the timer initialization on the distance to the critical point. *set_timer(distances)* computes a value based on the distance parameter.

A scale factor is also necessary, in order to normalize the interval of the possible values of the timer. Two characteristics of the algorithm are influenced by the scaling factor. The first is the overall propagation time of the location process. The second is the number of overlapping characteristics. The choice of a scaling factor is thus a tradeoff between the performances measured by the two characteristics. If the scale factor is to large, the propagation delay is also large and if the scaling factor is too small, the number of overlapping transmissions becomes too large.

4   Simulations

In this section we numerically validate the expected behaviour and performances of our algorithm. The simulations we present compare our localisation algorithm and the greedy localisation algorithm which is considered to be the reference. To compare algorithms, the criteria we are intersted in are (a) weather the localisation algorithm succeeds (are sensors localized at the end of the algorithm?) (b) the total number of emmitting sensors (c) the running time of the algorithm. The numerical experiments we present show that our algorithm (emission inhibition) competes with the greedy algorithm in terms of localisation success while significantly reducing the total number of emmiting sensors. Experiments also show that the inhibition *with selection* algorithm has more chances of making the localisation process succeed than the algorithm using inhibition, but *without* selection. In the case where collisions are considered, slowing down the localisation process by making sensors wait a long time before emitting their position is an obvious way of reducing the number of collisions, however this also augments the time required for the localisation process to finish. The simulations we present show that a waiting time which is increasing like the square of the critical distance, see Figure 1(c) seems to be perticularily appropriate and has a slowly increasing running time, which is an unexpected and pleasant property.

4.1   Details on the Experiments and the Representation of Results

An experiment consists in randomly dropping $n = 1000$ sensors in a 1 by 1 square. Three localised sensors are dropped in the middle of the square, and the localisation starts: each sensor with at least three localised neighbours (who emitted their localisation) becomes localised and broadcasts its position to all his neighbours (according to the algorithm which is being tested). The process goes on iteratively. In order to avoid the impact of the border effects we stop the simulation as soon as 500 sensor nodes are localised, and consider the localization process *succesfull*). Each experiment is repeated 2000 times, and the outcomes are presented in a box plot graphic.

Box plots are composed of a box, the lower line being the lower quartile, the middle one the median and the upper one is the upper quartile of the sample. The median is surrounded by a notch wich shows the 95% confidence interval (for the median). When comparing two medians, they are considered to be signicantly different only if one is not in the confidence interval of the other (i.e. if one median is not in the notch around the other median). The dashed lines extending above and below the box show the span of the other samples. The plus sign represents outliers. Statistically, outliers represent data which is suspected to be unsignificant, perhaps resulting from an input data error or bad measurement. However, in our setting the presence of outliers shows that we are in the transition phase between the regime where the localisation algorithms succeeds and the regime where it fails and the change of regime is not yet significant enough to be included in the statistical box plots.

4.2   Results

We present two sets of experiments. For the first set of experiments, collisions due to simultaneous transmissions are not taken into consideration. We are interested in observing how small the transmission radius can become while still allowing the localisation algorithm to succeed, as well as the total number of emmiting sensors. This set of experiments allows us to validate the inhibition algorithms (with and without selection) as significantly reducing the number of emissions while still competing with the greedy algorithm in terms of succeeding in localisating sensors.

The second set of experiment takes collisions into account. This set of experiments allows to compare the total time needed for the differents algorithms to achieve localisation. Intuitively, when a sensor is localised, it waits a certain time before broadcasting its position. The longer the waiting time, the less collisions their will be (which increases the probability that the localisation algorithm succeeds), but on the other hand, a longer waiting time will make the overall localisation process slower. We investigate the tradeoff between high sucess probability and fast localisation. The waiting time before retransmission is controlled by a parameter k. We find out that setting a waiting time quadratic in the distance to the critical region ensures that the inhibited with selection algorithm runs fast, independently of the scaling parameter k.

Numerical Experiments Without Collisions. The first set of experiments is composed of figures 2, 3 and 4. We look at the impact of the transmission range on the performance of the localisation algorithms. The x-axis of every figures corresponds to the range of emission which varies from 0.11 to 0.015 with a step of 0.005. With 1000 sensors in a 1 by 1 square, this corresponds to an expected number of neighbors ranging from 38 to 0.7 with a step of approximatively 2.5. On the left of each figure is the number of localised sensors versus the emission range. Because we stopped the simulations as soon as 500 sensors are localised,

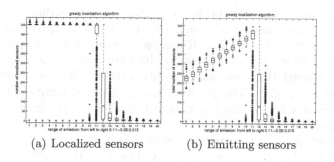

(a) Localized sensors   (b) Emitting sensors

Fig. 2. Greedy strategy

having 500 localised sensor means that the localisation algorithm has succeded. As this number deacreses the algorithm fails. On the right side of the Figures, we plot the total number of emission which occured at the time we stopped the simulations versus the emission range.

Looking at the Figures 2(a) and 3, we observe that the greedy algorithm is the more robust in the sense that it succeeds with a smaller transmission range than our algorithm (figure 3(a): emission inhibition with selection. The performances of our algorithm starts to decrease with an emission range of about 0.07 and the greedy algorithm with 0.05. This is expected, since with the greedy algorithm every localised sensor emits and this behaviour is necessarily more succeessfull than the inhibited algorithms (at least when collisions are not considered). Also, as expected, we observe on the right hand side of these Figures that our algorithm runs with far less emissions than the greedy one (in the case of the largest emission range about 120 emissions against 230). We also notice that for the smallest ranges, when the greedy algorithm is the only one succeding in localising sensors, the number of emissions is very large: nearly all the sensors have to participate to ensure that the localisation process succeeds. In this configuration, our algorithm inhibits some transmissions which were necessary to ensure the success of the localisation process: this explains why our algorithm stops working before the greedy algorithm and it suggests that that no localisation (trilateration) algorithm could make the localisation process succeed while significantly reducing the total number of emmissions (i.e., it would be running an "almost" greedy algorithm). Comparing Figure 3(a) and Figure 3(b) we notice that preventing the emission of sensors in a selective way (a sensor stops its timer only if the received data comes from a sensor located closer to the intersection of the 2 neighbours covarage circles) has an important impact. Indeed, numerical experiments confirm that the selective inhibition avoids to prevent a sensor to emit without the preventing emitting data covering the region it was expected to cover.

Comparing Figure 3(b) and Figure 3(c) confirms that choosing a waiting time increasing with the distance to the critical point of the sensor (c.f. definition 3) results in better performances than having a random waiting time (in terms of seeing the localisation algorithm succed even for small transmission ranges).

This comes from the fact that, in the case of the random waiting time algorithm, sensors situated far from the *2Coverage* region (c.f. Figure 1(b)) sometimes emit befor sensor closer to the critical *2Coverage* region. As a consequence, for the random waiting time, some of the sensors which are close to the *2Coverage* region no not emit although this would have permitted the localisation of the neighbours in the *2Coverage* regions. On the other hand, the total number of emission is smaller for the random waiting time algorithm than for the inhibition without selection algorithm because of the larger number of inhibitions. This strategy seems to be adequate as the range of emission is quite large.

As a first conclusion, these experiments show that the *inhibition with selection algorithm* significantly reduces the total number of emissions required to make the localisation process succed and that *it is optimal amongst inhibition algorithms* in the following sense: when the inhibition with selection algorithm starts to fail, the greedy algorithm requires almost every sensor to transmit.We also observe that when the expected number of neighbours (or equivalently when the transmission range or the density of the nodes) increases, the inhibition algorithm *without* selection and the inhibition algorithm with random waiting time, (c.f. Figure 4(b), 4(c)) compete in making the localisation process succeed. Moreover, when they significantly reduce the total number of emissions, even when compared to the inhibition with selection algorithm.

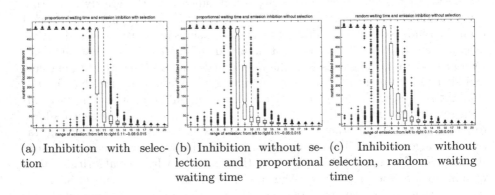

(a) Inhibition with selection

(b) Inhibition without selection and proportional waiting time

(c) Inhibition without selection, random waiting time

Fig. 3. Localized sensors

Numerical Experiments with Collisions. The second set of experiments deals with collisions and the impact of the waiting time function on the total time required for the localisation process to finish. We limit our study to the inhibition with selection localisation algorithm and compare it to the greedy algorithm. Also, we fix the transmission radius to r=0.11. We compare two waiting time functions,

1) Poportionnal waiting time $f(k) = kd$,

2) Quadratic waiting time $f(k) = kd^2$.

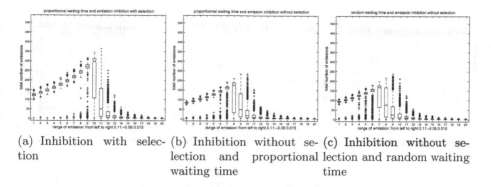

(a) Inhibition with selection

(b) Inhibition without selection and proportional waiting time

(c) Inhibition without selection and random waiting time

Fig. 4. Emitting sensors

We consider a normalised distance d which is the distance of a sensor to its critical intersection point divided by the transmission range of sensors, and we introduce a normalized parameter k. When a sensor is localised, it waits for a time $f(k)$ and broadcast its position (unless it becomes inhibited by a transmission from another sensor). Collisions are modelled in the following way: time is divided in discrete rounds, and if a sensor receives more than one message in the same round, it drops all of them . Intuitively a quadratic waiting time seems more appropriate since it is proportional to the area covered by the region and hence, on the expected number of sensors in the region.

Numerical results are plotted on Figures 5, 6, on the left side of the figures is plotted the number of localised sensors versus the parameter k which ranges from 201 to 6 with step 5. On the second plot of each of the Figures, we represented the total number of time before the simulation stops (either because 500 sensors are localised and the algorithm has succeeded, or because the localisation process fails). Notice that in the three Figures it can be seen that the total number of localized sensors often exceeds 500. This follows from the short waiting time which implies many simultaneous emission and the way the timeout criteria is implemented in our simulations.

Since the greedy algorithm implies that more sensor emit their position (c.f. section 4.2), it could have been that collisions have more impact on the greedy algorithm that the inhibition with selection algorithm. However, in terms of success of the localisation process, simulations shows that this is not the case: the greedy algorithm is again the more robust and numerical results show that it works in localizing sensors even when the waiting time parameter k becomes smaller.

On the contrary, the quadratic waiting time function makes the inhibition with selection localisation algorithm fail for the smallest values of k (figure 5(c)), a patological behaviour which is not as important when a proportional waiting time is used (c.f figure 5(b)). However, the important factor (once the localisation process succeds) is not to minimize k, but rather to minimize the *total running time of the localisation process*. With respect to this objectives

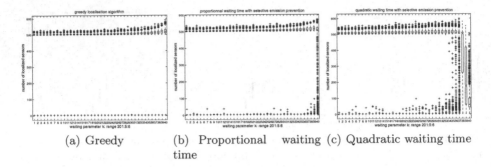

|(a) Greedy|(b) Proportional waiting time|(c) Quadratic waiting time|

Fig. 5. Localized sensors

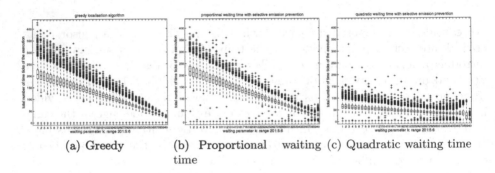

|(a) Greedy|(b) Proportional waiting time|(c) Quadratic waiting time|

Fig. 6. Execution time

(successful and fast localisation), simulations are to be interpreted as showing that the quadratic waiting time is the best.

Indeed, the running times of the greedy algorithm and the algorithm with proportionnal waiting time function are very similar (compare Figures 6(a) and Figure 6(b)) and even for small values of k, these running time are significantly greater than the running time of the algorithm with the quadratic waiting time for the choice of a greater k (c.f. figure 6(c)). Moreover, an important characteristic of the results obtained with the quadratic waiting time function is that the total running time is less sensitive to the value of the parameter k than the two others algorithms.

5   Conclusion

The numerical experiments conducted and presented in this paper show that the scheduling strategy introduced in this paper combined with emission inhibition with selection is good in minimizing the total number of emission as well as in reducing the total running time of the algorithm. Possible applications of this strategy would be to reduce the energy consumption by reducing the number of emissions, reducing privacy disclosure in the case security is a concern and reducing the total running time of the localisation phase. Intuitively, the strength of this algorithm might increase with the density of the sensors since the emission inhibition becomes more efficient. A possible future research direction would be to investigate theoretically the behaviour of our algorithm, including the impact of density.

References

1. Luo, H., Ye, F., Cheng, J., Lu, S., Zhang, L.: Ttdd: A two-tier data dissemination model for large-scale wireless sensor networks (2003)
2. Intanagonwiwat, C., Govindan, R., Estrin, D.: Directed diffusion: a scalable and robust communication paradigm for sensor networks. In: Mobile Computing and Networking. (2000) 56–67
3. Powell, O., Leone, P., Rolim, J.: Energy optimal data propagation in wireless sensor networks. arXiv.org automated e-print archives (2005) Report CS-0508052, journal version submitted for publication.
4. Efthymiou, C., Nikoletseas, S., Rolim, J.: Energy balanced data propagation in wireless sensor networks. (Invited paper in the Wireless Networks (WINET, Kluwer Academic Publishers) Journal, Special Issue on "Best papers of the 4th Workshop on Algorithms for Wireless, Mobile, Ad Hoc and Sensor Networks (WMAN 2004)")
5. Leone, P., Nikoletseas, S., Rolim, J.: An adaptive blind algorithm for energy balanced data propagation in wireless sensor networks. In: The First International Conference on Distributed Computing in Sensor Systems (DCOSS). Number 3560 in Lecture Notes in Computer Science, Springer Verlag (2005)
6. Savvides, A., Han, C.C., Strivastava, M.B.: Dynamic fine-grained localization in ad-hoc networks of sensors. In: Mobile Computing and Networking. (2001) 166–179
7. Savarese, C., Rabay, J., Langendoen, K.: Robust positioning algorithms for distributed ad-hoc wireless sensor networks usenix technical annual conference (2002)
8. Aspnes, J., Eren, T., Goldenberg, D.K., Morse, A.S., Whiteley, W., Yang, Y.R., Anderson, B.D.O., Belhumeur, P.N.: A theory of network localization. Unpublished manuscript. To appear, IEEE Transactions on Mobile Computing (2006)
9. Bulusu, N., Heidemann, J., Estrin, D.: Gps-less low cost outdoor localization for very small devices (2000)
10. Niculescu, D., Nath, B.: Ad hoc positioning system (aps) using aoa (2003)

The Threshold Behaviour of the Fixed Radius Random Graph Model and Applications to the Key Management Problem of Sensor Networks[*]

V. Liagkou[1,2], E. Makri[4], P. Spirakis[1,2], and Y.C. Stamatiou[2,3]

[1] University of Patras, Department of computer Engineering, 26500, Rio, Patras, Greece
[2] Mathematics Department, 451 10, Ioannina, Greece
istamat@cc.uoi.gr
[3] Research and Academic Computer Technology Institute, N. Kazantzaki, University of Patras, 26500, Rio, Patras, Greece
[4] University of the Aegean, Department of Mathematics, 83000, Karlovassi, Samos, Greece
liagkou@cti.gr, effiem@aegean.gr, spirakis@cti.gr, istamat@cc.uoi.gr

Abstract. In this paper we study the threshold behavior of the fixed radius random graph model and its applications to the key management problem of sensor networks and, generally, for mobile ad-hoc networks. We show that this random graph model can realistically model the placement of nodes within a certain region and their interaction/sensing capabilities (i.e. transmission range, light sensing sensitivity etc.). We also show that this model can be used to define key sets for the network nodes that satisfy a number of good properties, allowing to set up secure communication with each other depending on randomly created sets of keys related to their current location. Our work hopes to inaugurate a study of key management schemes whose properties are related to properties of an appropriate random graph model and, thus, use the rich theory developed in the random graph literature in order to transfer "good" properties of the graph model to the key sets of the nodes.

1 Introduction

As sensor networks are now a reality and they are ubiquitous in every aspect of everyday life (e.g. wireless mobile appliances, hot-spots, WLANs etc.) they have attracted the attention of many researchers from a variety of disciplines.

Briefly, a sensor network is a self-organizing, structureless, autonomous system of nodes with sensing and communication capabilities which move around freely within a region. Security is also a key concern in such networks due to the dynamic requests of previous unknown nodes for entering or leaving the network,

---

[*] Partially supported by the IST Programme of the European Union under contact number IST-2005-15964 (AEOLUS) and the INTAS Programme under contract with Ref. No 04-77-7173 (Data Flow Systems: Algorithms and Complexity (DFS-AC)).

S. Nikoletseas and J.D.P. Rolim (Eds.): ALGOSENSORS 2006, LNCS 4240, pp. 130–139, 2006.
© Springer-Verlag Berlin Heidelberg 2006

which raises trust issues amongst them. The use of wireless links also poses a major security issue in that there is less effort involved in compromising a communication session than that involved in compromising wired links. This, coupled with computational and energy constraints of the devices which comprise a sensor network, render the issue of security a most taxing problem. In this paper we focus specifically on the use of symmetric key cryptosystems for sensor networks, since the computational and the communication overhead incurred with public key schemes renders them most inefficient for these types of networks. Furthermore, the use of symmetric key cryptosystems requires that all nodes in the network share a common key for encryption and decryption.

Vital issues of security, such as authentication, integrity and confidentiality are dealt with using a symmetric key scheme; however the lack of a superior key management framework renders the whole security of the system ineffective since attacks are mainly made on the key management infrastructure of a system.

According to [6], there are three classes of key management. Namely: (i) key distribution, (ii) key agreement, and (iii) key pre-distribution. The authors of [6] propose a fourth class of key management: (iv) *distributed, key pre-distribution schemes.* Two good reviews on the key management problem and key predistribution schemes are [5] and [4]. In this paper we will propose a scheme which is based on properties of the *fixed radius random graph model* or *geometric graph model* (see, e.g., [7]). According to this model, nodes are drawn at random within a certain region and two nodes are considered connected (i.e. there exists an link between them) only if their distance is at most equal to some predefined value. By varying the distance as a function of the number of nodes, we obtain interesting behavior for the set of nodes as a whole (e.g. connectivity, triangle-freeness etc.). We prove that this model has a threshold behavior using techniques that were used to prove threshold behavior in the usual random graph models (see [11]), obtaining results similar to the ones obtained in [7] but avoiding a lengthy combinatorial analysis. We then propose a key distribution scheme according to which each node selects, dynamically, key sets depending on its position (which helps establish shared keys with nearby nodes since their key sets will be related due to the node's proximity). We believe that such a scheme is interesting since the properties of the key sets are directly related with properties of the random graph model, to which there is a rich mathematical theory to apply.

2 The Theoretical Tools

2.1 Random Graph Models

In this subsection we will define the basic random graph models that are currently used to model networks as graphs G whose vertices represent network nodes and edges a direct connection between pairs of nodes. In what follows, by n we will denote the number of network nodes and by Ω the set of all possible $\binom{n}{2}$ edges between these nodes.

- Model $\mathcal{G}_{m,m}$: select the m edges of G by selecting them uniformly at random, independently of one another from Ω.

- Model $\mathcal{G}_{n,p}$: include each edge of Ω in G independently of the others and with probability p.
- Model $\mathcal{G}_{n,R_0,d}$: generate n points in some d-dimensional metric space uniformly at random and draw an edge between two points only if their distance is at most R_0.

There is also another very useful graph model, called the *scale-free graph model* (see [10] for definitions and results related to this model) which is found to accurately model real, fixed topology networks (see for a good survey on these networks). This model, however, cannot model structureless networks (such as MANETs) and we will not refer to it further in this paper.

With regard to the model $\mathcal{G}_{n,R_0,d}$, in practical situations the metric space is either the 2-dimensional or 3-dimensional Euclidean space. In these spaces, distance is most often given through the l_2 Euclidean norm.

2.2   First Order Graph Properties and Thresholds

We are interesting in proving that certain properties related to MANETS exhibit a threshold behavior. We will be confined to properties expressible in the *first order language* of graphs. This language can be used to describe some useful (and naturally occurring in applications) graph properties using elements of the first order logic.

The alphabet of the first order language of graphs consists of the following (see, e.g., [11]):

- Infinite number of variable symbols, e.g. $z, w, y \ldots$ which represent graph vertices.
- The binary relations "$==$" (equality between graph vertices) and "$\sim$" (adjacency of graph vertices) which can relate only variable symbols, e.g. "$x \sim y$" means that the graph vertices represented by the variable symbols x, y are adjacent.
- Universal, $\exists$, and existential, $\forall$, quantifiers (applied only to *singletons* of variable symbols).
- The Boolean connectives used in propositional logic, i.e. $\vee, \wedge, \neg, \Longrightarrow$.

An example of graph property expressible in the first order language of graphs is the existence of a triangle: $\exists x \exists y \exists w (x \sim y) \wedge (y \sim w) \wedge (w \sim x)$. Another property is that the diameter of the graph is at most 2 (can be easily written for any fixed value k instead of 2): $\forall x \forall y [x = y \vee x \sim y \vee \exists w (x \sim w \wedge w \sim y)]$. However, other equally important graph properties, like connectivity, cannot be expressed in this language.

We will now define the important *extension statement* in natural language, although it clearly can be written using the first order language of graphs (see [11] for the details):

Definition 1 (Extension statement $A_{r,s}$). *The extension statement $A_{r,s}$, for given values of r, s, states that for all distinct $x_1, x_2, \ldots, x_r$ and $y_1, y_2, \ldots, y_s$ there exists distinct z adjacent to all x_is but no y_j.*

The importance of the extension statement $A_{r,s}$ lies in the following. When applied to the first order language of graphs, if $A_{r,s}$ (for *all* r, s) holds for a random graph G (in some random graph model) with probability tending to 1 asymptotically with the number of vertices of the graph, then for every statement A written in the first order language of graphs either $\lim_{n\to\infty} \mathbf{Pr}[G(n,p) \text{ has } A] = 0$ or $\lim_{n\to\infty} \mathbf{Pr}[G(n,p) \text{ has } A] = 1$.

In Section 3 we will prove, using results related to the distance between randomly chosen points in Euclidean spaces which we state in 2.3, that the extension statement $A_{r,s}$ holds, almost certainly, for all r, s for a random graph according to the fixed radius model.

2.3   Random Points in Euclidean Spaces

With regard to the distribution of the distance between points chosen uniformly at random to lie within a Euclidean sphere, the following was proved in [12]:

Theorem 1. *The probability density function and cumulative distribution function for the distance x between two random points within an d-dimensional Euclidean ball of radius R are given, respectively, by the following equations:*

$$P_d(s) = \frac{s^{d-1} \int_{\frac{s}{2}}^{R} (R^2 - x^2)^{\frac{d}{2} - \frac{1}{2}}}{\frac{1}{2d} B(\frac{d}{2} + \frac{1}{2}, \frac{1}{2}) R^{2d}} \tag{1}$$

$$D_d(x) = \int_0^s P_d(s)ds =$$
$$\left(\frac{x}{R}\right)^d - \frac{B_\alpha(\frac{1}{2}, \frac{d}{2} + \frac{1}{2})}{B(\frac{d}{2} + \frac{1}{2}, \frac{1}{2})} \left(\frac{x}{R}\right)^d + 2^d \frac{B_\alpha(\frac{d}{2} + \frac{1}{2}, \frac{d}{2} + \frac{1}{2})}{B(\frac{1}{2}, \frac{d}{2} + \frac{1}{2})} \tag{2}$$

with $0 \leq x \leq 2R$, $\alpha = \frac{1}{4}(\frac{x}{R})^2$.

The function $B_\alpha(x, y)$ is the *incomplete beta function*:

$$B_\alpha(x, y) = \int_0^\alpha t^{x-1}(1-t)^{y-1} dt$$

while $B(x, y)$ is the beta function which is equal to $B_\alpha(x, y)$ for $\alpha = 1$. For more on these function see, e.g., [1].

For the 2-dimensional case, which is of interest for *surface* sensor networks, we have the following:

Corollary 1. *For the circle (the 2-dimensional Euclidean ball) of radius R it holds that*

$$P_2(s) = \frac{2s}{R^2} - \frac{s^2}{\pi R^4} \sqrt{4r^2 - s^2} - \frac{4s}{\pi R^2} \arcsin \frac{s}{2R}, \text{ and}$$

$$D_2(x) = 4\left(\frac{x}{2R}\right)^2 + \frac{1}{2\pi} \left(\frac{x}{2R}\right) \left(4 - 4\left(\frac{x}{2R}\right)^2\right)^{\frac{3}{2}} - \frac{3}{\pi} \left(\frac{x}{2R}\right) \sqrt{4 - 4\left(\frac{x}{2R}\right)^2} +$$
$$\frac{2}{\pi} \arcsin \frac{x}{2R} - \frac{8}{\pi} \left(\frac{x}{2R}\right)^2 \arcsin \frac{x}{2R}. \tag{3}$$

3   Threshold Behavior of the Fixed Radius Random Graph Model

We will now study the threshold behaviour of the fixed radius random graph model with regard to properties expressible in the first order language of graphs.

Lemma 1. *For the 2-dimensional sphere (circle) the probability that $A_{s,t}$ fails for $\mathcal{G}_{n,R_0,d}$ is bounded from above as follows:*

$$\mathbf{Pr}[A_{s,t} \text{ fails in } \mathcal{G}_{n,R_0,2}] \leq \binom{n}{s+t} \left[1 - D_2(R_0)^s (1 - D_2(R_0))^t\right]^{n-(s+t)}. \quad (4)$$

Theorem 2. *If $\sigma = \frac{R_0}{2R} = c$ is a constant, $0 < c < 1$, then Equation (4) tends to 0. If $\sigma = \frac{R_0}{2R} = f(n) = \omega(\frac{1}{\sqrt{n}})$, then Equation (4) also tends to 0.*

Proof. From Equation (4), it follows that

$$\mathbf{Pr}[A_{s,t} \text{ fails in } \mathcal{G}_{n,R_0,2}] \leq$$
$$\binom{n}{s+t} \exp\left[-D_2(R_0)^s (1 - D_2(R_0))^t (n - (s+t))\right]. \quad (5)$$

Our goal is to find a condition on c such that the right-hand side of (5) tends to 0. Then $\mathbf{Pr}[A_{s,t}$ fails in $G(n, R, 2)]$ tends to 0 and, consequently, $\mathbf{Pr}[A_{s,t}$ holds in $\mathcal{G}_{n,R_0,2}]$ tends to 1 establishing the fact that any first order property holds, asymptotically, in $\mathcal{G}_{n,R_0,2}$ with probability 1 or 0.

Case 1. Let σ be a constant c, $0 < c < 1$. Then $D_2(R_0)$ is a constant too. Thus, the exponential factor of the right-hand side of Equation (5)

$$\exp\left[-D_2(R_0)^s (1 - D_2(R_0))^t (n - (s+t))\right] \quad (6)$$

tends to 0, for fixed s, t and n tending to infinity. Therefore, the probability $\mathbf{Pr}[A_{s,t}$ fails in $\mathcal{G}_{n,R_0,2}]$ also tends to 0.

Case 2. Let, now, $\sigma = f(n) < 1$, a function of n tending to 0. Then using power series analysis around 0, we obtain from (3) the following:

$$D_2(R_0) = 4\sigma^2 + \frac{1}{2\pi}\sigma(4 - 4\sigma)^{\frac{3}{2}} - \frac{3}{\pi}\sigma\sqrt{4 - 4\sigma^2} + \frac{2}{\pi}\arcsin\sigma - \frac{8}{\pi}\sigma^2\arcsin\sigma$$
$$= 4\sigma^2 - \frac{32}{3\pi}\sigma^3 + \frac{16}{15\pi}\sigma^5 + O(\sigma^6). \quad (7)$$

The term $D_2(R_0)^s (1 - D_2(R_0))^t$ in the exponent in (6) can be approximated as follows:

$$D_2(R_0)^s(1 - D_2(R_0))^t = 4\,s\,\sigma^2 - \frac{32\,s}{3\,\pi}\,\sigma^3 - [16\,s\,t + 8\,s\,(s - 1)]\,\sigma^4$$

$$+ \left[\frac{256\,s\,t}{3\,\pi} + \frac{16\,s}{15\,\pi} + \frac{128\,s\,(s - 1)}{3\,\pi}\right]\,\sigma^5 + O(\sigma^6) \quad (8)$$

with s, t constants. Then, from (6) and (8), it follows that if $\sigma = f(n) = \omega(\frac{1}{\sqrt{n}})$, then (6) tends to 0, for any s, t, completing the proof. □

The generalization, now, follows readily:

Theorem 3. *Let* $\sigma = \frac{R_0}{2R} = c$ *be a constant,* $0 < c < 1$. *Then for any first order property* A, *then* $\mathbf{Pr}[\mathcal{G}_{n, R_0, d}$ *has* $A]$ *tends to 1 or 0. If* $\sigma = \frac{R_0}{2R} = f(n) = \omega(\frac{1}{\sqrt[d]{n}})$, *then* $\mathbf{Pr}[\mathcal{G}_{n, R_0, d}$ *has* $A]$ *tends to 1 or 0 too.*

Although the property of forming a connected graph cannot be described in the first order theory of graphs, in [8] it is shown that for slighter larger values of σ, the network is almost certainly connected. More specifically, we only need to increase the threshold probability (in the 2-dimensional case) from $\frac{1}{\sqrt{n}}$ to $\frac{\sqrt{\log(n)}}{\sqrt{n}}$ to, also, ascertain connectivity in the resulting graph.

4   A Key Management Scheme Based on the Structure Induced by the Fixed Radius Model

We will now define a key management scheme which creates and discards, dynamically, key sets for sensor nodes depending on their current position. We will first explain the basic idea behind this scheme and then we will describe it in mode detail.

Generally, in key predistribution schemes, each network node is assigned a predefined set of keys before its entrance into a sensor network. As soon as the node enters the network, it may communicate securely with other nodes which carry sets of keys with non empty intersection with its own key set. If the node needs to communicate with nodes whose key sets do not intersect with its own, then the communication can be done in multiple hops via other nodes, much like in the routing problem in fixed topology networks). However, there are two problems with this approach. The first one is that one node may never need to communicate with nodes whose predefined key sets intersect its own. Similarly, one node may need to communicate more often with nodes with whom it shares no key under the specific key predistribution scheme. On top on these issues is also the fact that secure communication requires, first, plain communication which, in turn, requires node proximity, i.e. each node being within some certain range around the other.

Based on these considerations, we will propose a key management scheme which does not rely on predistribution but rather creates (and destroys) key sets dynamically for each node taking into account its physical position so as to form an interdependence between the key sets of physically nearby nodes and, thus, help these node to reach agreement on the keys which will be used for their communication.

For the details now, assume we have n nodes randomly distributed within a circle of radius R. We first fix a value C which, for each node, will define a circle centered at the node within which candidate keys will be considered. The radius C also models the communication range of each of the n nodes, meaning that their communication devices have sufficient power to transmit only within distance C away from the node (circle of radius C).

Assuming that the nodes are placed uniformly at random within the area of radius R, a fixed radius random graph with n nodes is formed so as to include edges between nodes only if their distance is at most $2C$ (i.e. their communication ranges/circles intersect). We also assume that each of the nodes knows its coordinates (e.g. using a GPS facility or some internal representation of the area in which the nodes move).

We now consider a discretization, a lattice, of the area with radius R which is known to the nodes. Thus, each of the nodes will occupy a point of the lattice. We are interested in estimating the number of lattice points lying within a radius C from a given node (we disregard minor discretization discrepancies since, asymptotically, they do not affect the estimates we will use). This is actually the problem, known as the *Gauss circle problem*, which asks for an estimate for the number of points within distance C of a given lattice point. This estimate is given below:

$$N(C) = \pi C^2 + E(C) \tag{9}$$

with $E(C) \leq 2\sqrt{2}\pi C$. See [13] for these values and very informative illustrations related to the Gauss circle problem as well as the more formal exposition of [2].

The combined (x, y) coordinates of these lattice points (which can easily be systematically produced by each node with only information its current position on the lattice) can form a set of keys to be used while interacting for establishing secure communication with its nearby neighbors within distance $2C$ (which are actually within physical communication reach). The number of shared keys can be computed as the number of points within the common part of two intersecting circles whose centers are at a distance s, which is given by the following equation (see [9]):

$$N = 2C^2 \arccos\left(\frac{s}{2C}\right) - \frac{1}{2}s\sqrt{4C^2 - s^2}.$$

Setting $s = \alpha C, 0 < \alpha \leq 1$ a constant, for the distance between the nodes, we see that the number of shared keys is equal to

$$C^2 \left[2\arccos\frac{\alpha}{2} - \frac{\alpha\sqrt{4 - \alpha^2}}{2}\right]$$

which gives each pair of nodes $\Theta(C^2)$ shared keys to choose from.

In addition, the dynamic key distribution scheme we described above actually frees us from the requirement to form a connected key set intersection graph since when two nodes come together, they will have formed the proper key sets (based on their current location) that will enable them to select common keys for their private communication.

We will now relate the key distribution scheme described above to threshold properties of the fixed radius graph model. First, with regard to the relationship between R, the range within which the nodes are moving, C, the transmission range, and n, the number of nodes we can draw some useful conclusions from the consideration of Theorem 2. In this context, $R_0 = 2C$, that is two nodes are considered adjacent if they can communicate (their ranges intersect), which occurs if their distance is at most $2C$). Thus, $\sigma = \frac{C}{R}$. Let $C = C(n)$ and $R = R(n)$ be functions of n tending to infinity and set $\sigma(n) = \frac{C(n)}{R(n)} = o(1)$. The assumption of $R(n)$ and $C(n)$ tending to infinity reflects the fact that as more nodes appear within a range, we should allow them to move in a wider area and, also, increase their communication range. The assumption $\sigma(n) = o(1)$, however, reflects the fact that we should not allow the nodes to increase the communication range too much, compared with the region within which they move, since the power dissipation will be excessive while, in addition, interference problems will appear. In addition, there would be an increased risk with nodes being able to receive signals destined to other nodes. This is reflected by having $\sigma(n) = \frac{C(n)}{R(N)} - o(1)$. Consider, now, possible ranges for $\sigma(n)$. According to Theorem 2, if $\sigma] = \omega(\frac{1}{\sqrt{n}})$ then the extension property holds with probability approaching 1 as the number of nodes increases. This means that all properties expressible in the first order language of graphs hold (asymptotically with n) either with probability 1 or 0. Especially, properties that are *monotonically increasing* (i.e. the probability of the property holding increases with increasing $\sigma(n)$) hold with probability 1 while their complementary properties hold with probability 0. What we need to do next is to define *good* properties, with regard to the key distribution scheme described above, which can be expressed in this graph language.

Let us consider the following property: *every two vertices have a common neighbor*. If this property holds, then for *each* pair of nodes there exists another node which shares keys with both of them. This may cause problems since it reduces the candidate keys which can be potentially used by the two nodes for the establishment of secure communication since the nodes should avoid the selection of keys which are also shared with the third node of the triangle. Setting $\sigma(n) = \frac{C(n)}{R(n)} = O(\frac{1}{\sqrt{n}})$, and since this property is monotone increasing, it holds with probability tending to 0. Thus its complementary property, which is a "good" property, holds with probability 1.

Other "good" properties could be the following:

- For any node v, its key set A_v is not a subset of the key set of any other node. Note that his property holds for the fixed radius random graph model and the key management scheme we introduced above.
- For any node v, its key set A_v cannot be a subset of the union of the key sets of l or less than l other nodes. Although this property cannot, possibly, be expressible in the first order language of graphs, it nevertheless can be

"approximated" by a property that is expressible: *no l nodes of the graph are adjacent, simultaneously, to any given node.*

5 Conclusion

In this paper we have examined the threshold behaviour of the fixed radius random graph model and its relation to the key management problem for sensor networks. We showed the using natural properties of this model we can define a dynamic key set creation scheme which takes into account the coordinates of the position of network nodes. We also demonstrated that a number of good properties can be described in the first order language of graphs and, thus, ensure that they hold with probability 1 as the number of nodes increases (e.g. the non-existence of a triangle) for certain ranges of the fixed radius random graph model.

We believe that the fixed radius random graph model has a large potential for the study of many properties related to the movement and secure communication of network nodes of a sensor network.

References

1. M. Abramowitz and I.E. Stegun, Eds., *Handbook of Mathematical Functions.* U.S. Department of Commerce, National Bureau of Standards, Washington, DC, 1972.
2. G.E. Andrews, *Number Theory*, W.B Saunders Company, 1971. Also by Dover Publications, 1994.
3. B. Bollobás, *Random Graphs*, Second Edition, Cambridge University Press, 2001.
4. S.A. Çamtepe and B. Yener, "Key distribution mechanisms for Wireless Sensor Networks: a Survey," Technical Report TR-05-07, Rensselaer Polytechnic Institute, Computer Science Department.
5. H. Chan, A. Perrig, and D. Song, "Random Key Predistribution Schemes for Sensor Networks", Proceedings of the IEEE Symposium of Privacy and Security", 11–14 May, pp. 197–213, 2003.
6. A. C-F. Chan and E.S. Rogers Sr., "Distributed Symmetric Key Management for Mobile Ad-hoc Networks", INFOCOM 2004. 23rd Annual Conference of the IEEE Computer and Communications Societies, pp. 2414–2424, 2004.
7. A. Goel, S. Rai, and B. Krishnamachari, "Monotone properties of random geometric graphs have sharp thresholds," Manuscript.
8. P. Gupta and P.R. Kumar, "Critical power for asymptotic connectivity," in *Proc. of Conf. on Decision and Control, Tampa, USA*, 1998.
9. J.W. Harris and H. Stocker, "Segment of a Circle," Section 3.8.6 in *Handbook of Mathematics and Computational Science*. New York: Springer-Verlag, pp. 92–93, 1999.
10. L. Li, D. Alderson, R. Tanaka, J. Doyle, and W. Willinger, "Towards a Theory of Scale-Free Graphs: Definition, Properties, and Implications (Extended Version)," Technical Report Technical Report CIT-CDS-04-006, Engineering & Applied Sciences Division California Institute of Technology, USA.

11. J. Spencer, *The strange logic of Random Graphs*, Springer Verlag, 2001.
12. S.-J. Tu and E. Fischbach, "Random distance distribution for spherical objects: general theory and applications to physics," J. Phys. A: Math. Gen. **35:** 6557–6570, 2002.
13. E.W. Weisstein, "Circle lattice points," From *MathWorld*–A Wolfram Web Resource. http://mathworld.wolfram.com/CircleLatticePoints.html

Area Based Beaconless Reliable Broadcasting in Sensor Networks

Francisco Javier Ovalle-Martínez[1], Amiya Nayak[1], Ivan Stojmenović[1], Jean Carle[2], and David Simplot-Ryl[2]

[1] SITE, University of Ottawa, Ottawa, Ontario, Canada K1N 6N5
{fovalle,anayak,ivan}@site.uottawa.ca
[2] IRCICA/LIFL, INRIA futurs, University of Lille, France
{jean.carle,david.simplot}@lifl.fr

Abstract. We consider the broadcasting problem in sensor networks where the nodes have no prior knowledge of their neighborhood. That is, to preserve power and bandwidth, no beacons or 'hello' messages are sent. We describe several Area based Beaconless Broadcasting Algorithms (ABBAs). In 2D, upon receiving the packet, each node calculates the ratio P of its perimeter, along the circle of transmission radius, that is not covered by this and previous transmissions of the same packet. The node then sets or updates its timeout to be inversely proportional to P. We also consider an alternative random timeout function. If the perimeter becomes fully covered, the node cancels retransmissions. The protocol is reliable, assuming an ideal MAC layer. We also describe three 3D ABBAs, one of them being reliable, each with two choices of timeouts. These three protocols are based on covering three projections, covering particular points on intersection circles, and covering intersection points of three spheres. Our protocols are the first reliable broadcasting protocols, other than blind flooding. We compare 2D ABBAs with two other existing beaconless protocols, BPS and Geoflood, showing its superiority. We also consider a MAC layer with collisions and show that all of our methods still remain very robust, showing high delivery ratio during the broadcast.

1 Introduction

In this article we consider sensor networks with nodes that have the same and fixed transmission radius. Broadcasting is an important task used for paging, alarming, location updates, route discoveries or even routing in highly mobile environments. Despite of its broad use and advantages, an increase in the density produces contentions, collisions and redundancies that limit the scalability of blind flooding. There exist several localized broadcasting algorithms that use intelligent techniques to reduce communication overhead during the broadcast. However, most of these protocols assume 1-hop or 2-hop knowledge. In order to discover their neighbors, the nodes in the network use beacons. The use of

S. Nikoletseas and J.D.P. Rolim (Eds.): ALGOSENSORS 2006, LNCS 4240, pp. 140–151, 2006.
© Springer-Verlag Berlin Heidelberg 2006

such 'hello' messages increases with mobility and changes in nodes activity status, resulting in critical bandwidth and energy losses. To address this issue, beaconless broadcasting, more intelligent than flooding, has been considered. In these protocols, the nodes act based on information gathered only during the broadcasting process. Beaconless broadcasting reduces energy consumption and increases broadcasting accuracy, since the protocols based on neighbor knowledge may have outdated information in dynamic ad hoc and sensor networks. This article studies the design of beaconless broadcasting protocols. The protocols proposed here require position information of each node to be available, which is not necessary to run blind flooding.

The first beaconless broadcasting protocols (other than blind flooding) are probabilistic and location based schemes by Ni, Tseng, Chen and Sheu [9]. However, they do not have good performance, either leaving many nodes uncovered or requiring most nodes to retransmit, as shown in [11]. We identified two existing recent beaconless broadcasting protocols: Optimized Broadcast Protocol (BPS) by Durresi, Paruchuri, Iyengar and Kannan [3] and Geoflood by Arango, Degermark, Efrat and Pink [1]. The authors claimed that their algorithm minimizes the number of necessary transmissions and that it outperforms all the other variations of flooding. However, BPS and Geoflood lack reliability.

The principal objective of this work is to develop several beaconless broadcasting algorithms. We propose four ABBAs, one for 2D and three for 3D spaces. Upon receiving the first copy of a message, each node sets a timer. Before the timeout expires, the node may receive more copies of the same message. Every node has a transmission coverage, which is assumed to be a circle in 2D and a sphere in 3D. If a node A receives a message from different sources and these sources cover the transmission circle or the sphere of A, then the node A has no need to retransmit the message. This means that each of its possible neighbors has received the message already from one or more nodes that cover it. If the node is covered before the timeout expires, then it decides not to transmit and ends the timer. We simulated our proposed broadcast algorithms, measuring the number of transmissions done for different network densities. We implemented OFP, three versions of Geoflood and compared them with 2D-ABBA. We showed that OFP and Geoflood are not reliable algorithms. In order to become reliable, Geoflood have to divide its transmission area in six angular regions instead of four. In our tests OFP failed considerably. In connected networks with average degree 7, OFP failed to complete a broadcast in 97% of the tested networks. Geoflood, even with 4 angles division, had higher retransmission counts than ABBAs. 2D-ABBA with coverage based timeout used on average 40% less messages than the reliable version of Geoflood, while the random 2D-ABBA had on average 28% less messages. Compared to Geoflood with 4 and 5 angles division, the original 2D-ABBA had on average 12% and 34% less messages respectively. Among our 3D proposals, 3D-ABBA1, based on covering three 2D projections, had the best results in terms of number of messages to complete the broadcast. However, this version lacked reliability. The second best 3D version (3D-ABBA2) performed very close to 3D-ABBA1 and had 100% delivery rate during the tests, although this version in theory is also unreliable. 3D-ABBA3 is a reliable broadcasting protocol. It is based on a covering theorem. A sphere A is covered by other spheres if every intersection point of three spheres from the set, one of them being A, is located inside another sphere from the covering set. Any three spheres determine two such common intersection points to be covered.

Protocol versions that used random timeout function were very competitive to the original optimized timeout values. Therefore we believe that our protocols will preserve good performance if implemented under a realistic MAC layer such as in IEEE 802.11 based networks. Recall that IEEE 802.11 use random timeouts in its backoff scheme.

The remainder of this work is organized as follows. Section 2 presents the related work for the broadcasting problem. In section 3 we describe 2D-ABBA. Section 4 describes 3D-ABBAs. Section 5 compares OFP and Geoflood with 2D-ABBA. Section 6 evaluates all 3D-ABBas. Finally section 7 concludes this article. 2D ABBA has been originally proposed in summer 2003 by last three authors, and its one paragraph description was published in [10, 6, 8]. The first three authors developed generalizations to 3D ABBAs in spring of 2004. This article summarizes master thesis work of the first author, defended in summer 2005.

2   Literature Review

Ni, Tseng, Chen and Sheu [9] studied the broadcast storm problem. Several beaconless schemes (probabilistic, counter-based, distance-based and location-based,) were proposed in [9] to reduce redundant rebroadcasts and differentiate the timing of rebroadcasts to alleviate this problem. In the probabilistic scheme [9], each node rebroadcasts the first copy of a received message with a given probability p . In the counter-based scheme [9], each node rebroadcasts the message if and only if it received the message from less than C neighbors. In the distance-based scheme [9], the message is retransmitted if and only if the distance to each neighbor that already retransmitted the message is greater than D. In the location-based scheme [9], the message is retransmitted if and only if the additional area that can be covered if the node rebroadcasts the message (divided by the area of circle with transmission radius) is greater than the threshold A . A simplified version of the method is to rebroadcast the message if the node is not located inside the convex hull of neighboring nodes that already retransmitted the message. However, these methods are not reliable. Further, the experimental data [9, 11] indicate low saved rebroadcasts for a high reachability, compared to other methods. We therefore did not include them in our experiments.

Bergonovo et all [2] described applications of broadcasting in inter-vehicle communications on the highway, with large and variable number of mobile terminals. The protocol appears to be a variant of the neighbor elimination scheme [11], which requires the maintenance of the neighbors list. This may not be an easy task in a highly mobile environment, the solution should be beaconless, and with guaranteed delivery.

Durresi, Paruchuri Iyengar, and Kannan [3] proposed a beaconless flooding algorithm named Optimized Broadcast Protocol (BPS). It is based on a hexagonal tiling of a plane, with the transmission radius as the edge length of the hexagons. Each broadcast packet contains two location fields L1 and L2, in its header. Whenever a node transmits a packet, it sets L1 to the location of the node from which it received the packet and sets L2 to its own location. The source node S sets both L1 and L2 to its location and transmits the packet. Each receiving node B first determines if the packet can be discarded. A packet can be discarded if the node has transmitted the same packet or if a node which is very close has already transmitted this packet. If the packet is not discarded, B determines if the packet came from the source S. If the message came from the source node S. If yes, B finds the nearest vertex V of a hexagon with center coordinates at S and with (Sx + R, Sy) as one of its vertexes. It computes its distance l from V and then delays the rebroadcast by a delay l/R. If the message came from other node K, then B selects the nearest hexagon vertex, and the rebroadcast is delay by l/20*R. After the delay elapses, B determines if the packet can be discarded. If the packet cannot be discarded, B sets L1 to the location of the node from which it received the message and L2 to its own location, and transmits. The authors of BPS propose avoiding retransmissions by having node B keep track of its distance dB to the nearest node that has retransmitted the packet. If this distance is greater than a threshold value Th, then B will retransmit. BPS authors proposed a value Th = 0.4R to ensure high delivery ratio while keeping number of transmissions low. As observed by Kim and Maxemchuk [7], the protocol may repeat flooding if the neighbors do not agree on the choice of the node near to the common ideal point. Authors [7] introduced a stopping rule to prevent that, and applied this type of flooding for route discovery. A more serious problem with BPS is its lack of reliability. As we will show in our experimental results, OFP can fail even in highly dense connected networks.

Arango, Degermark, Efrat, and Pink [1] proposed Geoflood, another beaconless broadcasting protocol. The algorithm is based on setting timers before retransmitting the message. Each node A defines a Cartesian plane with its location as the origin. Upon receiving the first copy of a message, node A records the quadrant from where the message was sent and sets a timeout. Authors [1] proposed that receiving nodes furthest away from the local sender should select smallest holding times. So timeouts increase as the distance to the sender decreases. They also introduced a random component to avoid contention that could arise between nodes located at the same distance from the sender. If node A receives more copies of the same message, at least one from each of other quadrants, before the timeout expires, it does not retransmit. The authors [1] recognized that their algorithm could fail in some situations. They showed the worst-case scenario in which part of the transmission area was not covered even when a node received messages from all four quadrants. They claimed that this area was relatively small and that other nodes in high-density networks would likely cover it. However, our experiments do not support such optimism.

Geoflood divides the transmission area in four quadrants or in four angles. This division causes the unreliability of Geoflood. In order to ensure reliability, Geoflood needs to divide into six angles instead of four. However, six transmissions from neighbors may not be necessary to cancel a retransmission, even in some cases three would be enough. This will bring unnecessary transmissions leading to unnecessary use of energy.

Recently Heissenbuttel, Braun, Wachli, and Bernoulli [4, 5] proposed a beaconless broadcasting protocol called: Dynamic Delayed Broacdasting (DDS). The idea behind DDS is basically the same to the one behind 2D-ABBA for nodes in two dimensions with position information available. Instead of taking into account perimeters as 2D-ABBA does, DDS takes into account the area left to be covered when a transmission arrives and decide not to transmit if the area left to be covered is less than a threshold value th. Unfortunately the use of th, does not guarantee delivery of the messages. The authors only proposed a 2D beaconless protocol, while we show one 2D and three 3D more sophisticated beaconless protocols.

3   Area Based Beaconless Algorithm in 2D

3.1   Asumptions

We assume that every node in the network knows its geographic position. Our second assumption is that all the nodes have the same transmission radius and use omni-directional antennas. Finally, nodes do not know where the neighboring nodes are, so no hello messages are used.

3.2   Description of 2D-ABBA

Upon receiving the first copy of a message that is broadcast, each node sets a timer. Before the timer expires, the node may receive more copies of the same message. If a node A receives a message from different sources and these sources cover the transmission circle of A, then node A has no need to retransmit the message. This means that each of its possible neighbors has received the message already from one or more nodes that cover it. If the node is covered before the timeout expires, then the node decides not to transmit and ends the timer. However, as long as the node is uncovered, the timer will continue to run, and when the timer expires the node transmits. Since all circles are of the same radius R, and each node A is aware only of the intersections with circles whose center B is inside its own circle (that is, $|AB| \leq R$), the coverage criterion can be simplified. Instead of covering the whole circle centered at A, only the perimeter of that circle needs to be covered. Further, when this perimeter is intersected by several circles whose centers are inside A 's circle, there can be up to two segments on the perimeter that are not yet covered. This property further simplifies the implementation. We propose two possible timeout functions for 2D-ABBA. The first one is a function that increases when the total length of uncovered portions of the circles with transmission radius around the node decreases. The function chosen was timeout = *degreesCovered* . The variable *degreesCovered* refers to the angle in degrees of the circle that has been covered. The second function that we considered is a random function with values between 0 and 1. This broadcasting protocol is reliable independently on the selected timeout function.

4   Area Based Beaconless Broadcasting Algorithms in 3D

4.1   Assumptions

We consider the transmissions patterns as spheres. All spheres are of the same radius, and each node A is only aware of transmissions from neighbors at distance $\leq R$. Instead of covering the whole sphere centered at A, one can consider only covering its 3D perimeter (set of nodes at distance $= R$ from A). All versions described here refer to covering that 3D perimeter.

4.2   3D-ABBA1

The idea behind our 3D-ABBA1 protocol is to use the three projections planes XY, XZ, and YZ. By projecting the transmission spheres into the three planes, the nodes are able to apply 2D-ABBA in each plane. Node A will not transmit if its corresponding transmission circles in all three planes have been covered. This in turn is simplified to covering the corresponding 2D perimeters. Again we define two different timeout functions. The first one is directly proportional to the three perimeters (one for each plane) covered by other nodes. In other words we considered $timeout = \frac{degsCovXY + degsCovXZ + degsCovYZ}{3}$. The variables *degsCovXY*, *degsCovXZ*, *degsCovYZ* refer to the angles in degrees of the circles that have been covered in the three corresponding planes. The second function considered is a random function with values between 0 and 1.

4.3   3D-ABBA2

When a node A receives a transmission, a circumference of intersection is created. If after receiving several transmissions each of the created circumferences is covered by other transmission spheres then the transmission sphere of node A is completely covered by others and it will not transmit. We further simplify computationally this scheme, in two ways, leading to two protocols 3D-ABBA2 and 3D-ABBA3. They both consider certain points for coverage instead of the mentioned whole circumference. In 3D-ABBA2 variant, we propose to select six points creating a hexagon that resides in the intersection circle (circumference). Thus instead of trying to cover the whole circumference, we propose to cover six points that reside on it. We apply the following criterion at each node A. If, for every such intersection with other transmitting sphere B, each of six considered points *P1- P6* is located inside another transmitting sphere (different from A and B), then the considered sphere A is covered in full. More precisely, let

$(d,\ e,\ f)$ be the coordinates of an intersection point of spheres centered at A and B. There must exist another transmitting node C with center coordinates $(xi,\ yi,\ zi)$ such that $(d-xi)^2 + (e-yi)^2 + (f-zi)^2 < R^2$.Note that different selected points may be covered by different transmitting nodes. Therefore a receiving node keeps a list of all the selected points. Whenever a new transmission is received the following actions are performed. A) Some existing selected points are inside the new transmission sphere. These points are eliminated. If no point remains, the sphere is completely covered. B)The new sphere is considered, precisely the new 6 selected points. Each such point is tested whether it is inside another existing sphere (spheres of previous transmitters). Those that are inside another one are ignored. Those that are outside any existing sphere are entered into the list of points to be covered.

Estimating intersecting volumes might lead to a more precise timeout function, but the calculation is too complex to be really used. A timeout that only depends on the number of uncovered selected points will lead to simultaneous retransmissions, because of few discrete possible values for such counter. 3D-ABBA2 uses an additional parameter directly related to the volume covered by the transmission. That volume depends directly on the distance between the sender and the receiver. For this reason, we decided that each selected point has a weight $\frac{R}{d}$. The parameter d is the distance between the transmitter that created the point and the receiver, while R is the transmission radius. In conclusion, we propose a timeout function: $Timeout$ = (number of received transmissions) + (sum of the weights of the selected points still to be covered).

In all our experiments, 3D-ABBA2 was always reliable. However, this is not a proof that it will always be the case. Unfortunately, there exist pathological cases where 3D-ABBA2 may fail to deliver to all the nodes connected to the source.

4.4   3D-ABBA3

To design protocol 3D-ABBA3, we considered the intersections of three spheres (more precisely, their 3D perimeters) rather than two. One of these spheres is centered by node A making the retransmission decision. Thus the intersection points are on its 3D perimeter. The other two spheres are centered at two neighboring nodes B and C that transmitted the message already. The intersection (if it exists) consists of two points. If each such intersection point is located inside another transmitting sphere D, then the considered sphere is covered in full. Therefore each receiving node A keeps a list of all such uncovered intersection points. Whenever a new node D transmits a message in the neighborhood then the following events occur. A)Some existing intersection points are inside the sphere centered at D. These points are eliminated. B) The new sphere centered at D is considered, with the current fixed sphere centered at A, for intersections with any other transmitting sphere, to find new intersection points. Each such point is tested whether it is inside another transmitting sphere. Those that are inside are ignored. Those that are outside any existing sphere are entered into the list of intersection points to be covered. When a transmission from a node D

does not generate new intersection points then the receiving node A adds node D to a list of transmitters that do not generate intersection points. We named this list as list of 'singular' spheres. Every time a message is received, node A checks its list of singular spheres to see if these nodes now create intersection points with the last received transmission. When an intersection is created, the corresponding node is removed from the list of singular spheres and the generated points are inserted in the list of intersection points. If the timeout for node A expires and both lists (the list of intersection points and the list of singular spheres) are empty, then node A is fully covered, and decides not to transmit. We again propose two possible timeout functions. The first one involves using the intersections created by the transmitting nodes with the receiving nodes (transmission patterns). The second is a random timeout function with values between 0 and 1. The reliability of 3D-ABBA3 follows from the following theorem.

Theorem 1. *Suppose that node A received transmissions from nodes $C1$, $C2$, ..., Cm. Consider all the intersection points X of the 3D perimeters of the spheres centered at the three nodes: A, Ci and Cj. If there exists at least one such intersection point and every such intersection point X is located (strictly) inside at least one of the remaining spheres, centered at Ck then the sphere centered at A is fully covered by the spheres centered at $C1$, ..., Cm, and node A does not need to retransmit, without affecting the reliability of broadcasting.*

Proof. Note that all the considered intersection points and all the intersection circumferences (between the sphere centered at A and any of the spheres centered at Ci for any i) are located on the 3D perimeter of the sphere centered at A. The conditions of the theorem are then topologically equivalent to an analogous theorem for the plane, applied, for example, in [12] for sensor area coverage problems. Both theorems use circles as closed curves that are intersected, but the circles are located in the plane and on a 3D perimeter (sphere), respectively. The proof, in its simplified form, is by contradiction. Assume that the condition of the theorem is satisfied, but there is an area (on the considered 3D perimeter) that is still not covered. Let Y be a point in this area. Find the closest curve (circle) to Y from the set. This curve separates the covered and the uncovered regions. Follow this curve until another curve is met. The meeting point is an intersection point from the theorem. It is easy to see that this intersection point is not strictly inside other curve (circle), which is a contradiction.

5   Comparison of 2D-ABBA with BPS and Geoflood

We implemented BPS with a threshold value $Th = 0.4R$. For this comparison with 2D-ABBA we implemented three versions of Geoflood; one dividing the transmission area in four angles, other in five angles, and finally one with division into six angles (the reliable version of Geoflood). For the tests, we considered a network of $n = 500$ static nodes, randomly distributed over an area of 100 x 100. We generated a total of 100 connected graphs for each of the following network degrees: $d = 7, 8, 9, 10, 15, 20, 25, 30, 35, 40, 45, 50, 60, 70, 80, 90,$

100, 125. After creating each of the connected graphs we started a transmission from a randomly selected node, and we measured the following characteristics to broadcast a single message to the entire network using the original and the random 2D-ABBA, the beaconless version of OFP and the three mentioned versions of Geoflood: a) average number of transmissions, and b) average percentage of nodes receiving the message during the broadcast. We define internal node as the node whose transmitting area or volume is covered by its neighbor nodes. In other words, if before performing a broadcast the transmission pattern of a node A is covered by the transmission patterns of its neighbors, then node A may or may not retransmit during a broadcast; however, if the transmission pattern of a node B is not covered by the transmission patterns of its neighbors then node B must retransmit during a broadcast. Such type of nodes will be called as external nodes. The average percentage difference between the two versions of 2D-ABBA was 4.55%, making us believe that the random version is a viable option to be deployed in IEEE 802.11 networks. The difference between the original 2D-ABBA and BPS was 44.24%. Here besides BPS performing worst, it also suppresses transmissions from external nodes, reaffirming its unreliability. The difference between 2D-ABBA and Geoflood of 4, 5, and 6 angles was 6.83%, 20.74%, and 29.22% respectively.

For the transmitting external nodes the percentage was 100% for 2D-ABBA and for Geoflood of five and six angles. BPS causes that a considerable percentage of external nodes will not transmit, which can cause broadcast failures. In average BPS caused that 13.29% of the external nodes did not transmit during the broadcast. Even for high-connected networks (degree = 125) 30.41% of the external nodes did not transmit. These results support, that BPS can fail even in high-connected networks. In the case of Geoflood of four angles the percentage of external transmitting also was considerable. In average, 10.56% of external nodes did not transmit.

5.1 Impact of a Realistic MAC Layer

In order to quantify the benefits of our new protocols, we repeated our experiments on a more realistic MAC layer, which considers message collisions. We have added a contention window of size CW and a timeout for each node before it can send any message. Any node randomly picks up an integer value between 0 and CW. Then, a node can neither receive two messages at the same time nor receive a message while transmitting. We selected a contention window of size 32, in accordance to the IEEE 802.11 standard. 2D-ABBA performed better than Geoflood and BPS. This clearly shows great potential of 2D-ABBA to be deployed in IEEE 802.11 networks. Although BPS was the most unreliable algorithm, it presented less percentage of collisions. For low degree networks, the collisions contributed to the broadcast failure in all the algorithms. From degrees between 7 and 10, all algorithms failed in several occasions; however for higher degrees only BPS failed at least in one occasion.

6 Performance Evaluation of the 3D Area Based Beaconless Broadcasting Algorithms

We considered a network of $n = 500$ static nodes, randomly distributed over a volume of 100 x 100 x 100. In order to control the average node degree d, we sorted all $\frac{n(n-1)}{2}$ (potential) edges in the network by their length, in increasing order. The transmission radius R is equal to the $nd/2$-th edge in the sorted array. We generated a total of 100 connected graphs for each of the following network degrees: $d = 7, 8, 9, 10, 15, 20, 25, 30, 35, 40, 45, 50, 60, 70, 80, 90, 100$. We considered ideal MAC and physical layers.

After creating each of the connected graphs we started a transmission from a randomly selected node, and we measured the following characteristics to broadcast a single message to the entire network using 3D-ABBA1, 3D-ABBA2, and 3D-ABBA3: a) average number of transmissions, and b) average percentage of nodes receiving the message during the broadcast The percentage of nodes receiving the message during the broadcast was 100% in every test for all ABBAs, except for 3D-ABBA1 where we found 2 unsuccessful tests, and the average number of nodes that did not get the message during these failures was only one. This shows that although not all versions of ABBA are 100% reliable (as shown in previous sections), all versions performed extremely well on the reliability criterion. Among the 3D versions, 3D-ABBA1 performed the best in terms of number of messages needed to complete a broadcast. However, since its reliability is not 100% we cannot immediately conclude that 3D-ABBA1 is the best protocol. The second best version was 3D-ABBA2; however, this is again, in theory, an unreliable protocol. The difference between 3D-ABBA1 and 3D-ABBA3 was on average about 25% of the messages for the original timeout function and about 20% for the random timeouts. Between 3D-ABBA3 and 3D-ABBA2 the difference was on average about 3% of the messages for the original timeout function and for the random timeout versions the difference was about 5% of the messages. The original timeout function in 3D-ABBA1 had about 10% fewer retransmissions than the random timeout function in 3D-ABBA1. However, when we applied random timeouts to 3D-ABBA2, we surprisingly obtained a minor improvement. In case of 3D-ABBA3, the difference was about 2% in favor of the original timeout function.Comparing the percentage of transmissions from internal nodes, 3D-ABBA1, 3D-ABBA2, and 3D-ABBA3 had a difference of 5.68%, 0.34%, and 2.32% between its two timer functions respectively. The difference between 3D-ABBA1 and 3D-ABBA2 was 15.11% when original timers were used and of 9.77% when random timers were tested. 3D-ABBA2 and 3D-ABBA3 had differences of 1.47% and 3.45% respectively. When high degree networks were tested only 30% of the internal nodes transmitted in 3D-ABBA3. This shows that 3D-ABBA3 achieves considerable energy savings while guaranteeing the reliable completion of a broadcast. The average difference of the percentage of transmitting external nodes was 10.80% between 3D-ABBA1 and 3D-ABBA2 (original timer functions). When using random timers, the difference was of 12.88%. The average percentage of transmitting nodes (external) shows that although 3D-ABBA1 and 3D-ABBA2 are not reliable, they both guarantee that almost all external nodes will transmit. In 3D-ABBA1 the minimum percentage of transmitting external nodes was of 77% but in 3D-ABBA2 this minimum value was of 94%.

6.1   Impact of a Realistic MAC Layer in the 3D-ABBA's

As in the 2D algorithms we simulated the effect of the MAC layer using discrete random timeouts between 0 and 32 (as in IEEE 802.11). The percentage of internal transmitting increased considerably; however, we still achieved energy savings while we were able to complete a successful broadcast. This shows that our 3D proposals have potential to be deployed in IEEE 802.11 networks.

7   Conclusions, Future Work, and Open Ideas

Area based beaconless broadcasting algorithm (ABBA) was proposed as a broadcasting protocol for sensor networks. This algorithm has fewer assumptions than almost all broadcasting algorithms. Localized broadcasting algorithms normally assume the use of 'hello' messages to provide the list of k-hop neighbors. The use of such beacons can cause large communications overhead and delivery failures due to outdated information. Our protocols, find neighbors only when they are really needed, and rely on geographic location of the nodes. We presented four beaconless broadcasting algorithms. All versions were based in setting timeouts before retransmitting; these timeouts depended on the transmission area still to be covered. We also tested the algorithms by setting their timeouts using a discrete random function, resembling the IEEE 802.11 functionality and simulating the MAC layer impact in our algorithms.

Several beaconless routing protocols were proposed in [9] but they were not competitive in comparisons made in other articles. We propose 2D-ABBA as new beaconless broadcasting algorithm in 2D, and compare it with two recent such protocols, BPS and Geoflood. We showed that BPS is not the best broadcasting protocol since it frequently lacks reliability. We showed that Geoflood protocol, in order to be reliable, has to divide the transmission area into six angles instead of four. We implemented BPS and three versions of Geoflood in order to compare them with 2D-ABBA. 2D-ABBA had less message count than all versions of Geoflood and BPS.

The 3D-ABBA versions presented in this work are the first beaconless broadcasting algorithms proposed, other than blind flooding, to work in 3D networks. Although only one 3D-ABBA version (3D-ABBA3) is fully reliable, the other two 3D versions performed almost perfectly in the experimental tests. More precisely 3D-ABBA1 failed only in 2 tests, while 3D-ABBA2 was always reliable in our experiments. 3D-ABBA1 performed better than 3D-ABBA2, and 3D-ABBA3, in terms of the number of messages to complete a broadcast. We also measured the percentage of external and internal transmitting nodes during a broadcast. We obtained that BPS suppressed a considerable percentage of the external transmitting nodes, while in fact, the percentage of the external transmitting nodes has to be 100% to guarantee reliability. This 100% was the case for 2D-ABBA, for Geoflood of 5 and 6 angles, and for both versions of 3D-ABBA3.

We were able to obtain excellent results with discrete random timeout variants. We believe that these show a very good promise to deploy our beaconless algorithms in IEEE 802.11 networks. Finally, we note that, besides serving for broadcasting purposes, our algorithms could also be a viable option for routing purposes.

Acknowledgments. This article is supported by grants from NSERC, CONA-CyT (projecto 37017-A) and INRIA. The first author was supported by CONA-CyT and SEP scholarships during the development of this work.

References

1. J. Arango, M. Degermark, A. Efrat, S. Pink, An Efficient Flooding Algorithm for Ad-hoc Networks, Proceedings of the 2nd Workshop on Modeling and Optimizations in Mobile Ad Hoc and Wireless Networks (WiOpt 2004), March 2004.
2. F. Bergonovo, A. Capone, M. Cesana, L. Fratta, L. Coletti, L. Moretti, N. Riato, Inter-vehicles communication: A new frontier of ad hoc networking, Proc. Medhoc, Tunisia, June 2003.
3. A. Durresi, V.K. Paruchuri, S.S. Iyengar, R. Kannan, Optimized Broadcast Protocol for Sensor Networks, IEEE transactions on Computers, Volume 54, Issue 8, August 2005.
4. M. Heissenbttel, T. Braun, M. Wlchli, and T. Bernoulli, Broadcasting in Wireless Multihop Networks with the Dynamic Forwarding Delay Concept, Technical Report, IAM-04-010, University of Bern, Switzerland, December 2004.
5. M. Heissenbuttel, T. Braun, M. Wlchli, Thomas Bernoulli, Optimized Stateless Broadcasting in Wireless Multi-hop Networks, IEEE Infocom (Infocom 2006); Barcelona, Spain, April 23-29, 2006.
6. F. Ingelrest, D. Simplot-Ryl, I. Stojmenovic, Energy-efficient broadcasting in wireless mobile ad hoc networks, in Resource Management in Wireless Networking (Mihaela Cardei, Ionut Cardei and Ding-Zhu Du, eds.), Springer, 2005, 543-582.
7. D. Kim, N.F. Maxemchuk, A comparison of flooding and random routing in mobile ad hoc network, 3rd New York Metro Area Networking Workshop, Sept. 2003.
8. X.Y. Li and I. Stojmenovic, Broadcasting and topology control in wireless ad hoc networks, Chapter 11 in: Handbook of Algorithms for Wireless Networking and Mobile Computing, (A. Boukerche ed.), Chapman & Hall/CRC, 2006, 239-264.
9. S.Y. Ni, Y.C. Tseng, Y.S. Chen, J.P. Sheu, The broadcast storm problem in a mobile ad hoc network, Proc. MOBICOM, Seattle, Aug. 1999, 151-162.
10. D. Simplot-Ryl, I. Stojmenovic, and J. Wu, Energy efficient backbone construction, broadcasting, and area coverage in sensor networks, in Handbook of Sensor Networks: Algorithms and Architectures (I. Stojmenovic, ed.), Wiley, 2005, 343-379.
11. I. Stojmenovic, M. Seddigh, J. Zunic, Dominating sets and neighbor elimination based broadcasting algorithms in wireless networks, IEEE Trans. on Parallel and Distributed Systems, Vol. 13, No. 1, January 2002, 14-25.
12. H. Zhang and J. C. Hou, Maintaining sensing coverage and connectivity in large sensor networks, Ad Hoc & Sensor Wireless Networks: An International Journal, 1 (1-2), 89-124.

A Flexible Algorithm for Sensor Network Partitioning and Self-partitioning Problems*

Sandip Roy, Yan Wan, and Ali Saberi

Washington State University

Abstract. Motivated by the need for agent classification in sensor networking and autonomous vehicle control applications, we propose a flexible and distributed stochastic automaton-based network partitioning algorithm that is capable of finding the optimal k-way partition with respect to a broad range of cost functions, and given various constraints, in directed and weighted graphs. Specifically, we motivate the need for new algorithms for network partitioning and distributed (or self-) partitioning. We then review our stochastic automaton-based partitioning algorithm, and extend its use for network partitioning and self-partitioning problems. Finally, the application of the algorithm to mobile/sensor classification in ad hoc networks is pursued in detail, and other applications are briefly introduced.

Keywords: sensor classification, partitioning, stochastic automata.

1 Introduction

Networks of communicating agents—including sensor networks and autonomous-vehicle teams—require distributed algorithms for a variety of tasks, including data communication/routing, estimation/agreement, and pattern-formation control, among others (see e.g. [1] for an interesting overview). In this article, we put forth the perspective that algorithms for network *self-partitioning* or *self-classification*, i.e. algorithms using which a network's nodes can form groups so as to minimize a cost while communicating in a distributed manner, are needed. We further contend that partitioning algorithms for these communicating-agent networks—whether distributed or centralized—must be flexible, in the sense that the algorithms should permit minimization of complex and varied cost measures. With these motivations in mind, we develop a flexible algorithm for network partitioning and self-partitioning using a stochastic automaton known as the influence model [2].

Distributed algorithms for self-partitioning are valuable for various sensor networking and autonomous vehicle control applications. While there is a wide literature on graph partitioning (which derives primarily from parallel-processing

---

* Correspondence should be sent to the first author at sroy@eecs.wsu.edu. The first and third authors were partially supported by the National Science Foundation under Grant ECS-0528882 (Sensors), and the third author was also partially supported by the Office of Naval Research under Grant N000140310848.

S. Nikoletseas and J.D.P. Rolim (Eds.): ALGOSENSORS 2006, LNCS 4240, pp. 152–163, 2006.

applications, see [5] for an overview), partitioning tasks for these *communicating-agent networks* are novel in several respects:

1) The algorithms used often must be distributed, e.g. because of the high power cost of communicating with a central agent or the need for security. For the same reasons, sparsity of communication in use of the algorithm is also often a must. Further, algorithms that are scalable, i.e. ones in which the computational cost for each agent grows in a reasonable manner with the network size, are needed; distributed algorithms can permit scalability.

2) The cost to be minimized is often a complex or multivariate one (e.g., for sensor network applications, delay, power dissipation, and reliability may each play a role in the cost), and varies from one applciation to another. Thus, we require algorithms that are flexible with respect to the cost minimized. This contrasts with the bulk of the literature on partitioning (e.g. [5]). in which algorithms are designed for a particular cost, typically a min-cut cost or a min-cut cost with partition-size constraints.

3) Communicating-agent networks are commonly subject to topological changes, for instance due to the addition of an agent or the failure of a particular communication link. Thus, partitions of the network may need to be adapted rapidly and frequently, ideally with minimal communication.

These novel requirements have motivated us to develop a distributed and flexible algorithm for network partitioning/classification. Specifically, we develop an algorithm for network self-partitioning (i.e., distributed partitioning) that is based on a stochastic automaton known as the influence model. We first introduced the influence model-based partitioning algorithm in [6], with the aim of exposing its use as a computational tool for partitioning. Here, we enrich the influence model-based partitioning tool for application to sensor self-classification and other network partitioning problems. The basic premise for using the influence model (specifically the copying influence model) for partitioning graphs is that groups of sites in the model that are separated by weak influences tend to have different statuses, while sites interconnected by strong influences tend to form a cluster with a common status. Therefore, by associating influences with edge weights in a graph, allowing the influence model to run for some time, and then examining the statuses, we can identify a good partition quickly with respect to many typical cost functions.

We have recently become aware of the literature on distributed clustering, in which very simple algorithms are used cluster sensor network nodes, typically by associating each node with a clusterhead (see e.g. [7]). Our approach to distributed partitioning is aligned with this research, in that we also use simple local associations to self-partition network nodes. What our stochastic approach to association permits is the formation of clusters that are optimal with respect to a global and flexible cost, albeit at somewhat higher computational cost than the simple clustering algorithms.

The remainder of this article introduces the graph partitioning problem, discusses the influence model and its use as a partitioning tool, and pursues the application of the distributed partitioning tool for sensor classification.

2   The Graph Partitioning Problem

Since one feature of our influence model-based partitioning algorithm is its flexibility, we find it useful to briefly introduce the general partitioning (classification) problem. A much more detailed introduction to the partitioning problem, and the connection of our algorithm to the existing literature in partitioning, can be found in [6].

Broadly, a k-way partitioning algorithm is concerned with classifying the vertices of a graph into k disjoint subsets. Specifically, let us consider a graph with (finite) vertex-set V that has cardinality n. We associate a positive *mass* m_v with each vertex (node) $v \in V$. In addition to the vertices, our graph also comprises a set of positively-weighted, directed edges. That is, for each ordered pair of distinct vertices v_i, v_j, we associate a weight $w_{ij} \geq 0$, where $w_{ij} = 0$ indicates that a directed edge is not present while $w_{ij} > 0$ indicates a weighted edge.

The partitioning problem that we consider is to classify the n vertices into k disjoint, non-empty subsets so as to minimize a cost function, while possibly enforcing one or more constraints. The cost function and constraints are phrased in terms of the total masses of the subsets and the edge weights on cuts.

Formally, we define a k-**way partition** of a graph as a subdivision of the nodes of the graph into k disjoint, non-empty **subsets** (components) $S_1, \ldots, S_k$. We are interested in identifying a partition that minimizes a **cost function**

$$f(M(S_1), ..., M(S_k), W(S_1, S_2), ..., W(S_k, S_{k-1})),$$

where $M(S_i) \triangleq \sum_{i \in S_i} m_i$ is the **mass** of subset i, and $W(S_l, S_m) = \sum_{i \in S_l} \sum_{j \in S_m} w_{ij}$ is the **size of the cut** between subsets i and j. We seek to minimize the cost function over the class of partitions that, in general, satisfy a number of constraints of the following types:

- **Algebraic constraints.** These are of the form
  $g(M(S_1), ..., M(S_k), W(S_1, S_2), ..., W(S_k, S_{k-1})) = 0$.
- **Set inclusion constraints.** These have the form $v_i \in S_j$, i.e. particular vertices are constrained to lie in particular subsets. We often refer to a vertex that is constrained to lie in S_j as a **reference vertex** for subset j.

We use the notation $S_1^*, \ldots, S_k^*$ for a partition that minimizes the cost subject to the constraints, and refer to this partition as an **optimal solution** of the partitioning problem[1]. Our aim is to solve the partitioning problem in a distributed manner, i.e. so that only communications along the edges of the graph are needed in finding the optimal partition. To the best of our knowledge, there have been no other algorithms developed that achieve partitioning without any global perspective at all in the graph.

---

[1] We can in fact allow a far more general cost function, e.g. one that depends on dynamics defined on the graph. We adopt this form here for clarity in our explanation of why our algorithm is expected to work well.

Of special interest to us within this broad class are problems that are subject to set-inclusion constraints, i.e. subject to constraints that certain reference nodes are contained in each component. We refer to such problems as *k-way partitioning problems with reference nodes*. Partitioning with reference nodes is of interest to us for several reasons: 1) problems in several distributed applications—for instance, the problem of grouping sensor nodes with base stations for multicasting—have this form, 2) these problems are well-known to be NP-hard for $k \geq 3$ and hence still require development of good algorithms, and 3) our algorithm is naturally designed to address this problem and hence gives fast solutions.

3   Using the Influence Model for Distributed Partitioning

We briefly review the influence model, and then describe its use as a distributed partitioning tool. The use of the influence in graph partitioning was first introduced in [6]; here, we enrich the development with sensor network partitioning and self-partitioning tasks in mind.

3.1   The Copying Influence Model: A Brief Review

Our algorithm for partitioning is based on evolving a stochastic automaton model. Specifically, we map a graph to a dynamic *stochastic network model*—a model in which values or statuses associated with network nodes are updated based on interactions with neighboring nodes. The statuses associated with the nodes form patterns as they evolve with time; these patterns turn out to identify good partitions of the graph. Since the automaton is updated only through interactions of nodes with graphical neighbors, it permits partitioning in a decentralized manner. The automaton that we use for partitioning is an instance of the *influence model* [2], a stochastic network automaton with a special quasi-linear structure. In this section, we very briefly review the influence model. We refer the reader to [2] for a much more detailed development.

An influence model is a network of n nodes or vertices or *sites*, each of which takes one of a finite number of possible *statuses* at each discrete time-step. We use the notation $s_i[k]$ for the status of site i at time k. We refer to a snapshot of all the sites' statuses at time k as the *state* of the model at time k. The model is updated at each time-step according to the following two stages:

1. Each site i picks a site j as its *determining site* with probability d_{ij}.
2. Site i's next-status is then determined probabilistically based on the current status of the determining site j. That is, the next status is generated according to a probability vector, which is parameterized by the current status of the determining site.

We shall only be concerned with a special case of the influence model called the *copying influence model*, in which each site takes on the same number k of statuses (labeled $1, \ldots, k$ w.l.o.g.), and furthermore each site simply *copies* the status of its determining site at each time step. To reiterate, at each time-step in the copying influence model, each site i picks a neighbor j with probability d_{ij} and copies the current status of that neighbor.

The influence model and copying influence model are compelling as modeling and algorithmic tools because they have a special quasi-linear structure, which permits analysis of the statistics of sites' statuses using low-order recursions. We refer the reader to [2] for details.

3.2   Algorithm Description

We can use the copying influence model as a tool for solving the partitioning problem described in Section 2 under rather broad conditions. Furthermore, since the influence model update only requires interaction among graphical neighbors (in a sense that will be made precise shortly), the algorithm is essentially decentralized (though a bit further effort is needed to *stop* the algorithm in a decentralized manner). The combination of flexibility and decentralization makes the influence model-based algorithm applicable to a range of partitioning tasks, including those discussed in the introduction. Let us first outline the algorithm, and then fill in the details.

1. **Mapping.** We map the graph to a copying influence model, by associating large influences with strong interconnections in the graph, and weak influences with weak interconnections. We note that we can permit asymmetric interconnection strengths.

2. **Initialization and Recursion.** We choose the initial state for the copying influence model. Here, the status of each site identifies the subset of the corresponding node in the graph. The statuses of the sites are updated recursively according to the developed copying influence model, and hence a sequence of possible partitions of the graph are generated. We note that this is a distributed computation, in that each site updates its status using only local information (i.e. information from graphical neighbors). Thus, in cases where a group of nodes in a real distributed system must self-partition, the influence model recursion can be implemented using localized communications between agents in the network. In presenting and analyzing the recursion, we find it convenient to first consider the case of partitioning with reference nodes[2], and then address partitioning problems without reference nodes.

3. **Stopping.** The recursion is terminated based on cost evaluations for a centralized algorithm and by decreasing influence model probabilities in the decentralized case. The statuses of the influence model at the stopping time specify the chosen partition.

---

[2] For notational convenience, we focus on the case where there is one reference node per component, but our development can straightforwardly be generalized to cases where the number of partitions is different from the number of references.

Mapping. We map the graph to a copying influence model with k possible statuses, with the motivation that we can identify a sequence of partitions of the graph by updating the influence model. That is, our algorithm classifies (partitions) the vertices in the graph according to the statuses of the corresponding influence model sites at each time-step of the recursion. The first step toward building this partitioning algorithm is to map the graph to a copying influence model, in such a manner that the copying probabilities in the influence model reflect the branch weights. In particular, we associate an influence model site with each vertex in the graph. We then choose the copying probabilities (influences) as

$$d_{ij} = \begin{cases} \frac{\Delta w_{ji}}{m_i}, & i \neq j; \\ 1 - \Delta \sum_l \frac{w_{li}}{m_i}, & i = j, \end{cases} \tag{1}$$

where Δ is chosen such that $\Delta \leq \dfrac{1}{\max_i \sum_j \frac{w_{ji}}{m_i}}$. Thus, large weights are associated with large influences, and small weights are associated with small influences; moreover, a large mass (inertia) m_i incurs small influence from other sites on site i (and large influence from itself), and a small mass m_i incurs large influence from other sites on site i. In general, the parameter Δ should be chosen large enough to achieve a fast convergence rate. We have specified the upper bound to guarantee that all the influence model parameters are valid.

In many decentralized and centralized applications, we envision this mapping stage as being done *a priori* by a centralized authority, even when the partitioning itself must be done in a decentralized manner. For instance, when new sensors are added to an existing network, the network designer can perhaps pre-program information about the communication topology and strengths of interactions between the sensors. However, it is worth noting that the mapping to the influence model is in fact inherently decentralized (i.e., an agent associated with vertex i in the graph can compute the weights d_{ij} from the vertex's mass and the weights of edges to neighbors) except in one sense: the scaling parameter Δ is a global one. Noticing that the maximum allowed value for Δ depends on the total weights of edges out of nodes and node masses, we note that Δ can often be selected *a priori* based on some generic knowledge of the graph topology (for instance, knowledge of the maximum connectivity of any single node), when decentralized mapping is also required.

Initialization and Recursion. Let us first develop a recursion for k-way partitioning with reference nodes (specifically, with one reference node per partition). For the problem of k-way partitioning with reference nodes, we fix the k *reference sites* (the sites in the influence model corresponding to the reference nodes) with distinct statuses from 0 to $k - 1$, and choose the initial statuses of other sites arbitrarily. Here, in order to fix the reference sites' statuses, we need to make a slight modification to the influence model developed in Equation 1 such that reference site i always chooses itself as the determining site:

$$d_{ij} = \begin{cases} 0, & i \neq j; \\ 1, & i = j, \end{cases} \tag{2}$$

(In a distributed context, notice that we only require that the reference nodes know their own identities to implement this initialization).

To generate a good partition, we then update the copying influence model. The state at each time-step of the recursion identifies a partition of the graph: that is, we classify the nodes whose associated sites are in status i in subset S_i. We note that the partition identified at each time-step automatically satisfies the set inclusion constraints for k-way partitioning with reference nodes. We shall show that this recursion, which generates a random sequence of partitions, eventually finds (passes through) the optimal solution with probability 1 under broad assumptions, after sufficient time has passed. We note that the recursion is completely distributed, in the sense that each node can decide its own subset at each time-step solely from its graphical neighbors.

In practice, we must develop a methodology for stopping the algorithm. Below, we discuss distributed and centralized approaches for stopping. The stopping methodologies seek to select low-cost partitions, while checking possible algebraic constraints. We shall show that appropriate stopping criteria permit identification of the optimal solution under broad assumptions, with probability 1.

Conceptually, one might expect this partitioning algorithm to rapidly identify low-cost partitions, because strongly-connected sites in the influence model (sites that strongly influence each other) tend to adopt the same status through the influence model recursion, while weakly-connected sites do not influence each other and hence maintain different statuses. Recalling that the influence strengths reflect edge weights and node masses, we thus see that the partitions identified by the model typically have strongly-connected subsets with weak cuts between them. For many typical cost functions, the optimal partition comprises strongly-connected subsets with weak links, and hence we might expect the algorithm to find good cuts quickly.

For k-way partitioning (without reference nodes), we can find the optimum by solving the partitioning problem with reference nodes for all sets of distinct reference node selections, and optimizing over these. Such a search may be impractical when a large number of partitions is desired However, most applications of interest to us have natural reference vertices, so we do not focus on the case without references.

A few further notes about the recursion are worthwhile:

1) For simplicity of presentation, we have implicitly assumed that agents have a common clock. However, we can equivalently use an update in which each site changes its status at random times; this strategy is similarly tractable.

2) Regarding scalability in a distributed setting, we note that each agent in a network only needs to randomly select a neighbor and poll that neighbor at each time-step, so the processing/communication per time step does not increase with the size of the network. The total processing/communication cost thus scales with the duration of the recursion. The algorithm's duration scales well with the network size as compared to other partitioning algorithms, for problems with reference nodes.

3) In some applications, we may already have one partition of a graph, and may wish to improve on this partition (with respect to a cost of interest) or to adapt the partition to changes in the graph. In such cases, we can speed up the recursion by initializing the influence model based on the original partition.

Stopping. Again, consider the k-way partitioning problem with reference nodes. (The adaptation to the general k-way problem is trivial.) For centralized problems, the global partition is known to a central agency at each recursion stage (time-step) and hence the cost can be evaluated and constraints can be checked. The minimum cost partition found by the algorithm can be stored. In this case, we propose to stop the updating after a waiting time, i.e. when the minimum-cost partition has not changed for a certain number of algorithm stages. We will show that a sufficiently long waiting time guarantees that the optimal solution is identified.

For distributed problems, it is unrealistic that a single agency can evaluate the global cost of a partition as in the centralized case, since each node only has available local information. A simple strategy in the distributed case is the blind one. Another clever strategy for distributed stopping is to use an influence model with state-dependent parameters. In particular, we progressively isolate (reduce the influence) between sites with different statuses after each update (and increase the self-influence correspondingly), until the influence model is disconnected (partitioned). More specifically, for each update, the (time-varying) influence $d_{ij}[k]$ is modified as follows:

- If $s_i[k] \neq s_j[k]$ and $d_{ij}[k] \geq \delta$, then $d_{ij}[k+1] = d_{ij}[k] - \delta$ ($i \neq j$) and $d_{ii}[k+1] = d_{ii}[k] + \delta$;
- If $s_i[k] \neq s_j[k]$ and $d_{ij}[k] < \delta$, then $d_{ii}[k+1] = d_{ii}[k] + d_{ij}[k]$ and $d_{ij}[k+1] = 0$ ($i \neq j$);
- If $s_i[k] = s_j[k]$, then $d_{ij}[k+1]$ remains the same.

When this time-varying algorithm is used, we note that the statuses of sites converge asymptotically (see Figure 1). This is because the influence model becomes disconnected, so that each partitioned component has only one injecting site and is guaranteed to reach consensus. Thus, a partition is found asymptotically. Furthermore, it is reasonable that this algorithm finds a good partition, since weak edges in the original graph tend to have different statuses at their ends in the influence model, and hence these edges are removed by the algorithm. We refer to this strategy as **partitioning with adaptive stopping**.

3.3   Algorithm Analysis: Summary

In the interest of space, we omit entirely detailed performance analysis of the partitioning algorithm. This analysis, which is based on global characterization of the influence model, can be found in [6] and [8]. Here, let us simply summarize the main results of this analysis:

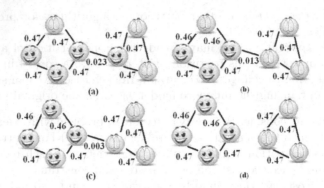

Fig. 1. This diagram illustrates how a network partitions itself (based on the update of the time-varying copying influence model) in a totally distributed manner

1) Under very broad connectivity conditions on the optimal partition, the recursion can be shown to pass through the optimal partition with probability 1. Hence, for a broad range of cost functions, k-way partitioning with and without reference nodes and with centralized stopping can be achieved.

2) Totally-distributed partitioning with adaptive stopping (and with reference nodes) can be shown to find the minimum-cut k-way partition (the partition with the minimum total edge weight across the cut edges) with probability 1, when the influence-reduction step size is chosen sufficiently small. Our simulations (omitted in the interest of space) indicate that totally-distributed partitioning can be achieved at moderate computational cost.

3) When the cost of the optimal partition is sufficiently small compared to the cost of other possible partitions in a partitioning problem with reference nodes, we can show that the optimal partitioning in polynomial time with respect to the size of the network.

4   Application: Sensor/Mobile Classification

Distributed partitioning holds promise as a tool for classification in distributed sensor networks and mobile ad hoc networks, e.g. for the purpose of multicasting or of transmitting information from the sensors/mobiles back to "leader nodes" or base stations or central authorities.

There is a wide literature on *routing* in ad hoc networks when the absolute positions of the sensors/mobiles are known (see [9] for a survey of methods). Recently, distributed algorithms (specifically local-averaging methods) have been used to infer location information in the case where absolute positions are unknown except at peripheral locations (see e.g. [10]), and hence permit development of routing algorithms for these networks. Beyond routing, classification of sensors/mobiles with base stations is an important task, for the purpose of multicasting (transmitting information to many destinations from multiple sources) or so that subsequently data can be routed to and from appropriate base stations to the sensors/mobiles.

Several recent articles have addressed multi-hop multicasting in ad hoc networks (see e.g. [3]). In multicasting applications as well as other settings where data may be transmitted to/from several sources or base stations, classification of mobiles/sensors with the base stations is important. We contend that the influence model-based partitioning tool can advance the state-of-the-art on classification in ad hoc networks, for several reasons:

1) As made clear by the comparison of location-known and location-unknown algorithms for routing, decentralized algorithms for classification may be needed in cases where there is no central authority with full knowledge of the network.

2) We may need to optimize the classification with respect to several (possibly complex) cost criteria (including for example minimum (or average) hops to each base station, average delay cost, and various reliability criteria). In fact, the costs may depend on the specifics of the decentralized algorithm used for routing/multicasting. The influence model-based algorithm permits us to consider multiple and complex cost criteria.

3) Often, the topologies of sensor networks and mobile ad hoc networks change with time, and hence it is beneficial to use an algorithm that can update the optimum with little effort (in either a distributed or centralized case). The influence model-based algorithm has this advantage.

For illustration of this application, we have used the influence model-based algorithm for sensor classification in a small example (one with 3 base stations and 27 sensor nodes). The example was generated by placing the 30 sensors in a uniform i.i.d. manner within the unit square, allowing communication between sensors within 0.3 units of each other, and choosing three sensors (Sensors 3, 14, and 20) to also serve as base stations. We associate a weighted undirected graph with the sensor network in which the 30 vertices correspond to the 30 sensors, and branches indicate communication between sensors. Each branch weight is chosen to be inversely proportional to the distance between the pair of sensors, with the motivation that longer communication links are more apt to failure and delay and hence are more weakly connected. We consider 3-way partitioning of this graph with reference vertices 3, 14, and 20 using the influence model algorithm.

We consider partitioning with centralized stopping, with respect to two cost functions:

1) We partition the graph so as to *maximize* the minimum of the positive eigenvalues of the *Laplacian matrices* associated with the three subsets (partitions)[3]. The minimum non-zero eigenvalue of the Laplacian associated with each subset is well-known to indicate the connectivity of that subset, and can be used to bound several relevant graph-theoretic properties such as the graph diameter (see [11] for a full development). By maximizing the minimum among

---

[3] We notice a maximization problem can routinely be converted to a minimization by choosing a cost that is the negative of the original cost.

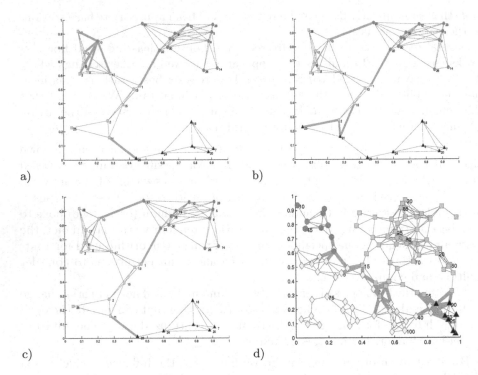

a)   b)

c)   d)

Fig. 2. Partitioning a 30-sensor network with reference nodes a) based on a minimum-subgraph-eigenvalue cost, b) based on a greedy-routing cost, and c) with distributed stopping. We also partition a 100-sensor network based on a minimum-subgraph-eigenvalue cost (d).

the non-zero eigenvalues, we thus find a partition with strongly-connected subsets and weak links between them. The optimal partition with respect to this *minimum-subgraph-eigenvalue* cost measure is shown in Figure 2.

2) Our second cost measure is motivated by consideration of low-cost and low-overhead distributed routing for ad hoc and sensor networks. A simple greedy algorithm for routing when location information is available is to send the message to the node (sensor) closest to the destination during each transmission (see e.g. [9]). Assuming such a greedy routing algorithm is used, our aim is to classify the sensors with base stations so that the maximum number of hops to a sensor from its base station is minimized. (The average number of hops could be used instead.) Thus, we partition the graph using this maximum number of hops when greedy routing is used. The optimal partition when this *greedy-routing cost measure* is shown in Figure 2. We note that, as expected, the optimal partition has subsets which are more balanced in size but contain weaker links, as compared to optimum for the minimum-subgraph-eigenvalue measure. This example highlights an interesting advantage of the influence model: the greedy-routing cost function does not admit an analytical form but can be computed for a given partition, but nevertheless the optimal partition can be found.

We also consider min-cut partitioning with distributed (adaptive) stopping for this example. The result is shown in Figure 2. We note that such a distributed algorithm could be implemented in the sensor network itself, and would only require individual sensors to have local parameters (in particular, distances to neighbors). Such a distributed algorithm might be especially useful in cases where the topology is subject to change, so that the sensors must re-classify themselves periodically.

As further illustration, we also show a 4-way partition with reference nodes of a 100-sensor network (based on a minimum-subgraph-eigenvalue cost) in Figure 2.

References

1. H. Qi, S. S. Iyengar, and K. Chakrabarty, "Distributed sensor networks – a review of recent research", *Journal of the Franklin Institute*, vol. 338, pp. 655-668, 2001.
2. C. Asavathiratham, S. Roy, B. C. Lesieutre and G. C. Verghese. "The influence model," *IEEE Control Systems Magazine*, Dec. 2001.
3. C. Cordiero, H. Gossain, and D. P. Agrawal, "Multicast over wireless mobile ad hoc networks: present and future directions," *IEEE Networks Magazine*, Jan./Feb. 2003.
4. S. Roy, A. Saberi, and K. Herlugson, "Formation and Alignment of Distributed Sensing Agents with Double-Integrator Dynamics", IEEE Press Monograph on *Sensor Network Operations*, 2004 (in press).
5. B. Chamberlain, "Graph partitioning algorithms for distributed workloads of parallel computations," *Technical Report UW-CSE-98-10-03*, University of Washington, Oct. 1998.
6. Y. Wan, S. Roy, A. Saberi, and B. Lesieutre, "A stochastic automaton-based algorithm for flexible and distributed network partitioning," *Proceedings of the 2005 IEEE Swarm Intelligence Symposium*, Jun. 2005.
7. F. Nocetti, J. Gonzalez, and I. Stojmenovic, "Connectivity based k-hop clustering in wireless networks," *Telecommunication Systems*, vol. 22, pp. 205-220, 2003.
8. "A flexible stochastic automaton-based algorithm for network self-partitioning," submitted to the *International Journal of Distributed Sensor Networks*.
9. M. Mauve, J. Widmer, and H. Hartenstein, "A survey of position-based routing in mobile ad hoc networks," *IEEE Networks Magazine*, vol. 6, pp. 30-39, Dec. 2001.
10. A. Jadbabaie, "On geographic routing with location information," submitted to *Proceedings of the IEEE Conference on Decision and Control*, The Bahamas, 2004.
11. F. R. K. Chung, *Spectral Graph Theory*, American Mathematical Society Press: Providence, RI, 1997.

Computing Bridges, Articulations, and 2-Connected Components in Wireless Sensor Networks

Volker Turau

Hamburg University of Technology, Institute of Telematics
Schwarzenbergstraße 95, 21073 Hamburg, Germany
turau@tuhh.de

Abstract. This paper presents a simple distributed algorithm to determine the bridges, articulation points, and 2-connected components in asynchronous networks with an *at least once* message delivery semantics in time $O(n)$ using at most $4m$ messages of length $O(\lg n)$. The algorithm does not assume a FIFO rule for message delivery. Previously known algorithms either use longer messages or need more time. The algorithm meets the requirements of wireless senor networks and can be applied in several areas relevant to this field such as topology control, clustering, localization and virtual backbone calculations.

1 Introduction

Sensor networks - networks of small, resource-constrained wireless devices embedded in the physical environment - present new challenges to the design and implementation of distributed algorithms. Energy efficiency is the key to prolonging the network life time and is thus of primary importance. Communication is the main consumer of energy and the consumption grows with the lengths of the messages. If energy consumption is not equally distributed over all nodes in the network, hot spots will emerge. This will lead to an early failure of these nodes, which may result in a disconnected network, that is no longer able to fulfill its task. Hence, it is important to identify hot spots and to use alternative routing paths to equally spread the load of communication.

The topology of a wireless sensor network is represented by an undirected graph $G = (V, E)$ where V is the set of nodes and $E \subseteq V \times V$ the set of edges describing the available communication links: (u, v) belongs to E means that u can send messages to v and vice versa. The topology depends on uncontrollable factors such as node mobility, weather interference noise, multi-path fading as well as on controllable parameters such as transmit power. Articulation points of G are likely to become hot spots, if they fail or run out of energy the network becomes disconnected. Hence, the number of articulation points of G reveals the number of weak points within the network topology. A high degree of fault-tolerance is achieved for networks that are k-connected, and this

S. Nikoletseas and J.D.P. Rolim (Eds.): ALGOSENSORS 2006, LNCS 4240, pp. 164–175, 2006.

property increases with increasing k. These networks provide multiple-path redundancy between every pair of nodes enabling techniques such as long sleeping periods and load balancing. Using random geometric graphs Bettstetter proved in [1] that the probability that a network with $n \gg 1$ nodes, each node with a transmission range r_0, and a homogeneous node density ρ is k-connected is $P(G$ is $k-$connected$) \approx P(\delta \geq k)$ for $P(\delta \geq k)$ almost one (δ denotes the minimum node degree of G) and this probability rises with ρ and r_0. Increasing the node density comes with higher costs and is very often not a valid option. The value of r_0 is of crucial importance for the functionality of the network. If the transmission power is too low, connectivity of all nodes is not guaranteed. If the power is too high, there may be too much interference, i. e., multi-user interference may not allow for efficient use of bandwidth. Furthermore, increasing the transmission range means a higher consumption of energy.

The topology of a wireless sensor network can be controlled by some tunable parameters such as transmitting power and antenna directions. This process is called topology control and extensive research has been done in this field in recent years [2,3,4,5,6]. The goal is to allow each node in the network to adjust its transmitting power individually (i. e., to determine its neighbors) so that a *good* network topology can be formed. In [2] Ramanathan and Hain propose a simple heuristic to achieve biconnectivity through the control of transmission power. If the network is connected but not biconnected each node attempts to do biconnectivity augmentation as follows. Every node sets a timer with a value t that is randomized around an exponential function of the distance from the next articulation point. If after time t the network is still not biconnected the node increases its power to the maximum possible. Nodes closer to an articulation point are more likely to remove the articulation and therefore given priority using timers. To determine the biconnected components a centralized algorithm is used, this is not a realistic option in non-static networks. The topology control algorithm presented in [6] extends the work of Ramanathan et al., but computing 2-connected components is still not distributed.

This paper contributes towards a solution for the topology control problem: A simple distributed algorithm to determine the bridges, articulation points, and 2-connected components in an asynchronous network using at most $4m$ messages with a length of $O(\lg n)$. Compared with the currently best distributed depth-first algorithms, this is only an increase of $n - 1$ messages, these are needed to inform each node about the 2-connected components they belong to. The assumptions for this algorithm meet the requirements of wireless sensor networks: The semantics of message delivery is *at least once* and the FIFO rule for message delivery is not assumed (i. e., messages transmitted over a link in the same direction can arrive at the other end of the link in any order). The footprint of the algorithm is small: On the average a node needs to store $2m/n + 2$ identifiers in addition to a few local variables. The output is available in a distributed manner: Each link knows whether it is a bridge, each node knows whether it is an articulation and is aware of the 2-connected components it belongs to. Previously known algorithms either use longer messages or need more time.

Many algorithms developed for wireless sensor networks can benefit from the knowledge of articulations points. Connected dominating sets (CDS) have been proposed as virtual backbones of wireless ad hoc networks [7]. All articulation points are included in every minimum CDS. Hence, identifying articulations points can reduce the communication effort for determining a CDS: If $M_1, \ldots, M_s$ are CDSs of the 2-connected components and A is the set of articulations of G, then $\bigcup_{i=1}^{s} M_i \cup A$ is a CDS of G (not necessarily a minimum CDS). T. Hara proposed a replica allocation method in ad hoc networks were the biconnected components form groups of nodes maintaining replicas [8]. Span is a power saving technique for multi-hop ad hoc wireless networks that reduces energy consumption without significantly diminishing the capacity or connectivity of the network [9]. When a region of a network has a sufficient density of nodes, only a small number of them need be on at any time to forward traffic. Obviously, putting nodes that are articulations into sleep mode leads to a disconnected network. Hence, this work also benefits from the knowledge of the articulations and biconnected components. Further applications domains include clustering (a cluster should be fully contained in a 2-connected component), localization of nodes ([10]), and TDMA slot assignment algorithms.

This paper is organized as follows: Section 2 summarizes related work and Section 3 defines the computational model used for this work. The algorithm and its analysis is presented in Section 4. A short discussion of an implementation of the algorithm using a real wireless sensor network concludes the paper.

2   Related Work

The presented algorithm is the first algorithm for this problem with time complexity $O(n)$ using $O(\lg n)$-messages. The messages consist of an identifier (3 bits) and at most one additional integer. The algorithm transmits $4m$ messages in the worst and $2m + n - 1$ messages in the best case, m being the number of links in the network. If message delivery is guaranteed in a single unit of time, the algorithm terminates within $2n + d - 2$ time units in the worst case (d is the depth of the search tree). Hohberg [11] presents a distributed algorithm for the problem at hand using $2m + n - 1$ messages. The proposed algorithm needs $2m + n - 1$ units of time and is consequently considerably slower than our algorithm. While all messages of Hohberg's algorithm have length $O(\lg n)$, in our algorithm $m - n$ of the messages have only length $O(1)$. In [12] Chaudhuri presents a distributed algorithm for the restricted problem of finding the bridge-connected components. While the number of messages is $O(n)$, it uses messages of length $O(n)$ and it is based on the FIFO rule. An algorithm for synchronous networks is described in [13]. The communication costs are not analyzed in detail, but are high compared with our algorithm. Swaminathan and Goldman present in [14] an incremental distributed algorithm for computing the 2-connected components in a dynamically changing graph. After a new edge is inserted it takes $O(b+c)$ messages of length at most $O(b(b+e))$ to recompute the 2-connected components, where c is the number of 2-connected components and

b and e are the numbers of nodes and edges in the resulting 2-connected component. The algorithm is rather complex and an implementation would probably overstrain the main memory of many of the currently available sensor nodes.

3 Computational Model

This paper assumes an asynchronous model based on message passing. Let $G(V, E)$ represent a connected communication network, with V being the set of nodes and E the set of bidirectional communication links. The asynchronous network has the following properties:

1. No two nodes in the network share memory.
2. The semantics of message delivering is *at least once*, i.e., there is a guarantee that every message sent is also received, but a node may receive a message more than once. The most straightforward realization of this semantics is through ARQ: Upon receiving a message, a node sends an acknowledgment to the sender. If a sender does not receive an acknowledgment within a given unit of time, the message is sent again until the receiver confirms the receipt of the message.
3. Message delivery times vary and cannot be predicted or bounded.
4. Messages sent over a link are not corrupted. This can be achieved by using error-correcting codes in combination with ARQ.
5. Every message reaches its receiver after a finite amount of time.
6. Messages sent over a link may not necessarily arrive in the same order they were sent (i.e., no FIFO rule for message delivery is assumed). This can be a consequence of the way messages are handled at the MAC-layer.
7. Every node is aware of all its links. A node knows the identity of the link over which a message is received. Nodes have a unique identifier and have no global knowledge about the network.

This computational model meets the requirements of most current wireless sensor networks. The main weakness of the model is that it does not tolerate failures of links or nodes. In case a node fails during execution, the algorithm terminates without computing all articulations.

4 Algorithm

The proposed algorithm is based on the distributed depth-first search algorithm invented by Cidon [15] and corrected by Tsin [16], which does not assume the FIFO rule. Interleaved into the determination of the depth-first tree is the calculation of the articulations using a technique from the sequential algorithm for this problem due to Tarjan [17]. The main challenge is to develop an algorithm within the computational model that keeps the total number of messages and their lengths as small as possible.

4.1   Informal Description of the Algorithm

The algorithm uses five types of messages and marks the links at each end with an attribute. Initially all link ends are marked UNVISITED. The messages Forward and Backtrack are used to explore the network in a depth-first order. The message Visited is used to avoid sending the message Forward to a node that is already discovered. Upon receiving the message Forward for the first time, a node sends concurrently to its normal course Visited messages to all neighbors except to the father of this node. Nodes receiving a Visited message mark their end of the link as VISITED. This attempts to prevent the neighbors from considering this node as yet undiscovered and from sending a Forward message to it at some later time. If however due to delays, a node v_1 is discovered before a Visited message from a previously visited neighbor v_2 has arrived, a Forward message might be sent from v_2 to v_1. To deal with this situation, node v_1 must discard the Forward message and v_2 must send a Forward message to another undiscovered neighbor (or a Backtrack message to the father node if all neighbors are already discovered). This behavior is enforced by the following two rules:

- An already discovered node ignores Forward messages received over links marked UNVISITED.
- A node receiving a Visited message over a link marked SON sends a Forward message to another undiscovered neighbor.

As shown by Tsin in [16] the unpredictability of message delivery times can cause other situations that must be taken care of to retain the depth-first character of the search. Due to these complications the classical algorithm of Tarjan to determine articulations and 2-connected components based on the so-called *low values* needs to be adapted. The low value of a node is the minimum of

1. the depth of the node,
2. the depth of the nodes reachable via back-edges from this node, and
3. the low values of the sons of the node.

The challenge is to extend the depth-first algorithm such that it computes the low values of all nodes without sending additional messages or using long messages. Compared with the algorithm of Tsin, only one additional type of message is needed (message Inform) and the length of some messages is increased to hold an additional integer. The message Forward carries the depth of the current node to a son. Upon receiving a Forward message for the first time, a node x determines its own depth. By broadcasting its depth with a Visited message, the neighbors of x get notified that x has been visited. This information is used to carry out step 2 for calculating low values. In case a Visited has already been received over a link, the node at the other end of the link has a lower depth, and the current node will not change its low value. Therefore, the depth is only included in Visited messages sent over links marked UNVISITED, an empty Visited message is still sent over links marked VISITED to retain the depth-first search. At any time a node stores the lowest of the depth values transmitted by

Visited messages. All but at most one of the Visited messages arriving at a node come over back edges. Consider the case where a node v_1 has two successors v_2, v_3 in a depth-first search tree. Once the node v_1 is discovered it sends a Forward message to v_2 and a Visited messages to v_3. But when the search returns to v_1 a Forward message is sent to v_3 (for an example consider nodes 3, 4, and 5 in Figure 1). Hence, the value carried by the Visited messages sent to v_3 must not be used for the calculation of the low value of node v_3. As proved in Section 4.3, this situation can easily be recognized. On receiving a Backtrack message carrying the low value of a son step 3 is accomplished. Finally the low value of a node is calculated immediately before the search backtracks to a father node.

A block is a maximal, non-trivial connected subgraph without an articulation, blocks are either 2-connected components or bridges. Articulations are recognized upon the receipt of a Backtrack message: According to [17], if the depth of a node is less than or equal to the low value of a son, then the node is an articulation. In case the depth is strictly less than the low value the corresponding edge is marked as a bridge otherwise it is marked CLOSED. This marks the *end* of a 2-connected component. To identify the nodes that belong to 2-connected component, the message Inform is used. Upon the detection of an articulation point an Inform message carrying the identifier of the new block is recursively sent over all reachable SON links except those marked CLOSED, i. e., over the edges of the depth first search tree of the current block. Inform messages are sent concurrently to the depth-first exploration of the rest of the network. This way, each node will know the block it belongs to. Block identifiers can be assembled from the id of the articulation together with the link identifier. Node identifiers are only needed for this calculation, in practical applications block identifiers can also be randomly generated. Inform messages are sent in parallel to exploring the rest of the graph. The opposite ends of bridges are recognized just before a Backtrack message is sent: If the low value is equal to the depth, the link is a bridge and it is marked accordingly.

4.2   Formal Description of the Algorithm

All nodes execute the same algorithm, which consists of a handler for every of the five types of messages. All nodes allocate memory for the following seven variables:

boolean *root* ← *articulation* ← **false**
boolean *state* ← UNDISCOVERED
int *low* ← *depth* ← MAX_VALUE
List *links*, *block_ids* ← ∅

The variable *root* indicates that the node is the root of the search tree. The variable *articulation* will hold the value **true** if and only if the node is an articulation. All nodes are initially in state UNDISCOVERED, upon receiving the first Forward message, the state changes into DISCOVERED. The list *links* stores the state of each link of the node, there are eight different states. Initially all links have state UNVISITED. The initially empty list *block_ids* will finally contain the identifiers of all blocks the node belongs to. All nodes implement the following two routines:

```
procedure SEARCH
    if ∃ link l s.t. l.state = UNVISITED then
        l.state ← SON
        send new Forward(depth) over l
    else if root = true then
        if |block_ids| = 1 then
            articulation ← false
        end if
    else
        l ← link in state FATHER
        low ← min(depth, low)
        if low = depth then
            l.state ← BRIDGE_FATHER
        end if
        send new Backtrack(low) over l
    end if
end procedure
```

```
procedure RESTART(link l)
    if l.state = UNVISITED then
        l.state ← VISITED
    else if l.state = SON then
        l.state ← VISITED
        SEARCH
    end if
end procedure
```

At every node the following operations are carried out when a message of the specified type is received over link l. It is assumed, that the nodes execute the operations corresponding to a message completely before executing the code of the next message, i.e., incoming messages are put into a queue of bounded length.

▷ Message *Backtrack(son_low)*

```
if l.state = SON then
    if depth <= son_low then
        if depth < son_low then
            l.state ← BRIDGE_SON
        else
            l.state ← CLOSED
        end if
        articulation ← true
        generate new block_id
        block_ids.insert(block_id)
        send new Inform(block_id) over l
    else
        l.state ← BACKTRACKED
    end if
    low ← min(son_low, low)
    SEARCH
end if
```

▷ Message *Forward(p_depth)*

```
if state = UNDISCOVERED then
    state ← DISCOVERED
    l.state ← FATHER
    depth ← p_depth + 1
    if low = p_depth then
        low ← depth
    end if
    SEARCH
    send new Visited(depth) over all
        links in state UNVISITED
    send new Visited() over all links
        in state VISITED
else
    RESTART(l)
end if
```

▷ Message *Start*

---

state ← DISCOVERED
depth ← *low* ← 0
root ← **true**
SEARCH
send new *Visited*(*depth*) over all links in state UNVISITED

▷ Message *Inform*(*block_id*)

---

if *block_id* ∉ *block_ids* **then**
    block_ids.insert(*block_id*)
        send new *Inform*(*block_id*) over all links in state BACKTRACKED
end if

▷ Messages *Visited*(*p_depth*) and *Visited*()

---

if *l.state* = UNVISITED ‖ *l.state* = SON **then**
    if *p_depth* included && *p_depth* < *low* **then**
        low ← *p_depth*
    end if
    RESTART(*l*)
end if

The procedure SEARCH implements the depth-first search, a node explores every possibility of extending the search via all links before backing up to the father node. The procedure RESTART is used to restart the search after a message has been received out of order due to message delays. The structure of the depth-first tree is available through the states of the links. Links marked FATHER or BRIDGE_FATHER lead to the predecessor of a node and links marked CLOSED, BRIDGE_SON or BACKTRACKED lead to successors. The ends of edges that are bridges are labeled BRIDGE_SON and BRIDGE_FATHER. The algorithm is invoked by sending the parameterless message Start to a node that will be used as the root of the search tree. Figure 1 demonstrates an application of the algorithm.

4.3   Correctness and Analysis

To prove the correctness of the algorithm the concept of a token traveling through the network is used. The token is introduced by the message Start. It is passed on by the messages Forward and Backtrack, but only in the following two cases:

1. Forward passes the token to the receiving node if this node is in state UNDISCOVERED.
2. Backtrack passes the token to the receiving node if the message is received over a link in state SON.

First, we prove that at every point in time there is a single token in the network. Every attempt to pass on the token is initiated by a call of the procedure SEARCH. It suffices to prove that whenever a node calls procedure SEARCH it

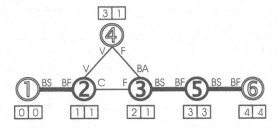

Fig. 1. Network with 6 nodes, bold nodes are articulations and bold edges are bridges, and node 1 is the root. The nodes are visited in ascending order of their identifiers. The nodes are annotated with the pair *depth, low* and the links are annotated with their final state. The blocks are $\{1, 2\}, \{2, 3, 4\}, \{3, 5\}$ and $\{5, 6\}$.

possesses the token and since the node last received the token the node has not passed it to another node. Initially only the root node possesses the token. If SEARCH is called upon receiving a `Forward` message at a node v_1, then either the node was in state UNDISCOVERED when the message arrived and thus just received the token, or it was in state DISCOVERED. In the latter case SEARCH is only called when the message was received from a node v_2 over a link l in state SON. Thus, a `Forward` message was previously sent over l (but in the opposite direction). Since the state of l is still SON, no `Backtrack` message was received over l. This implies that this is the link over which v_1 has sent its last `Forward` message. If this `Forward` message has passed the token to v_2, then v_2 must have been in state UNDISCOVERED and had not sent any message over this this link and also would never send a `Forward` message over that link. But since the link is in state SON, this is impossible. Hence, the last `Forward` message did not pass the token and v_1 is still in possession of the token.

SEARCH is also called upon receiving message `Visited` over a link marked SON. As in the previous case, the last `Forward` message did not pass the token and the node possesses the token at this time. The only other case SEARCH is called, is when a `Backtrack` message is received over a link in state SON. But then the node has just been given the token. This proves, that at every point in time there is a single token in the network and thus, no two branches are followed concurrently. Since every node explores every possibility of extending the search via all links before backing up to the father node, the search employs the depth-first order and the token traverses the network in depth-first order. This also proves that the algorithm terminates. Figure 2 depicts all possible transitions of states of ends of links. The states FATHER and BRIDGE_FATHER are called father-states and SON, CLOSED, BRIDGE_SON and BACKTRACKED are called son-states. Note that once an end of a link has a son- or father-state, this property holds forever with only one exception, it is possible to transit from SON to VISITED.

In the following we prove that the algorithm correctly labels all links, i. e., tree-edges with father- respectively son-states and back-edges with state VISITED. As shown above, the token traverses the network in depth-first order, i. e., the edges the token is passed on are exactly the tree-edges. A node receiving the token

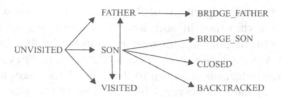

Fig. 2. Transitions of states of ends of links

for the first time - except the root node - marks that link with a father-state and it remains in a father-state. A node marks a link as SON before sending a Forward message. If this message passes on the token, it remains forever in a son-state. Otherwise, the node will later receive a Visited or Forward message that will change the label to VISITED. Furthermore, if a node in state DISCOVERED receives a Forward message over a link l, then l is already in state VISITED or will change into that state. This proves, that the algorithm correctly labels all links. Because of the idempotence of routine RESTART, receiving duplicate messages of type Visited or Forward is not a problem. On the first receipt of a message Backtrack the state of the link is changed, hence any Backtrack message not received over a link in state SON can be safely discarded. A similar argument proves that duplicate Inform messages also do no harm.

To prove the correctness of the algorithm it remains to prove, that the algorithm correctly computes the low values. First, the case of receiving a VISITED message over a tree-edge is considered. The depth value included in this message must not be used to calculate the low value of the receiving node. A correction is necessary, if and only if the depth value carried by a Forward message equals the current lowest depth value included in all Visited messages received so far. This only occurs if the Forward message was received over the same link as the Visited message leading to the current lowest depth value. In this case the current lowest depth value is equal to the depth of the father node. Since a low value of a node is bounded by the depth of the node, the error can be safely corrected by taking the depth of the node as the current lowest depth value, this is just the current lowest depth value incremented by 1.

When a node is ready to send a Backtrack message to its father, there are no links marked UNVISITED. Furthermore, the node will pass the token to the father node and the token will never return to this node. Hence, after this event the node receives no more Visited messages over links marked SON. Since Visited messages are never received over links marked VISITED or FATHER, the node receives no more VISITED messages at all. Thus, the node has received the depth of all neighbors reachable via back edges with Visited messages and has discarded VISITED messages received over tree-edges. Furthermore, the node knows its own depth and the low values of all sons from the corresponding Backtrack messages. Hence, the node is in a position to calculate its own low value and to include it in the Backtrack message to be sent to its father. On receiving a Backtrack message with the low value of the son, a node can decide whether it is an articulation.

The depth-first search algorithm of Tsin transmits at most $4m - (n - 1)$ messages. Our algorithm uses in addition the Inform messages, one per tree edge. Thus, the total number of messages is bounded by $4m$. In the best case, there are only two Visited messages per back edge and one Forward, Backtrack, and Inform per tree edge, summing up to $2m + n - 1$ messages in the best case. All but $m - (n - 1)$ messages carry 3 bits to identify the kind of the message and an integer less than n. Visited messages sent over links marked VISITED carry only the message identifier, there is at least one such message for every back edge. Thus, the message length is $O(\lg n)$. In the best case our algorithm transmits $m - (n - 1)$ integers less than Hohberg's algorithm, in the worst our algorithm transmits $m - (n - 1)$ more integers.

The time complexity is the maximum duration from sending the Start message until the termination of the algorithm under the assumption that the time of delivering a single message over each link is at most one unit of time. This assumption is only made for this calculation, the algorithm operates correctly for any finite message delivery time. In the following it is proved, that passing the token takes at most one unit of time. For this purpose a node that passes successfully the token with a Forward message at time t to a node v is considered. The node v must have been in state UNDISCOVERED when receiving the Forward message, this did not happen later than time $t + 1$. All Visited and Forward messages destined for node v were sent prior to time t, the time the father of v send the Forward message to v. Hence, at time $t + 1$ node v must have received all of these messages. Thus, v knows whether it has send a Forward message without passing the token. This yields, that at time $t + 1$ node v passes the token either to a son or back to its father. Since the token is passed twice along every tree edge, the depth first search needs $2n - 2$ units of time. In the worst case when the graph is 2-connected sending the Inform messages is started after the depth-first search has finished. In this case our algorithm needs d additional units of time (d is the depth of the search tree). In total the algorithm requires no more than $2n - 2 + d$ units of time. This is a considerable improvement over previous algorithms. Using induction, it is straightforward to prove that the sum of the lengths of the lists $block\_ids$ of all nodes is at most $2(n - 1)$. This bound is attained for trees only. Hence, on the average there are about two identifiers in each list. Furthermore, on the average each node has $2m/n$ neighbors, the state of each can be safely stored in one byte.

5   Conclusion

This paper presented a novel distributed algorithm for computing bridges, articulations, and 2-connected components of undirected graphs in time $O(n)$ using at most $4m$ messages of length $O(\lg n)$. In order to improve this result either a more efficient distributed depth-first search algorithm is needed or a completely different approach must be taken (i. e., not based on depth-first search). The algorithm is suitable for usage in many application of wireless sensor networks due to its short messages and its robustness, i. e., it does not assume a FIFO rule for message delivery and is immune from message duplication. The algorithm has been successfully implemented and tested on a real wireless sensor network consisting of ESB and ECR nodes of the ScatterWeb platform developed at the FU Berlin [18]. Each node was equipped with 2 kByte RAM and 64 kByte EEP-ROM and with a wireless communication device working at 19200 Bits/s. The experiment was conducted using 24 nodes with varying topologies.

References

1. Bettstetter, C.: On the minimum node degree and connectivity of a wireless multihop network. In: Proc. ACM MobiHoc, ACM (2002) 80–91
2. Ramanathan, R., Rosales-Hain, R.: Topology control of multihop wireless networks using transmit power adjustment. In: IEEE INFOCOM 2000. (2000) 404–413
3. Borbash, S., Jennings, E.: Distributed topology control algorithm for multihop wireless networks. In: Proc. Int. J. Conf. on Neural Networks. (2002) 355–360
4. Lloyd, E., Liu, R., Marathea, M.V., Ramanathan, R., Ravi, S.: Algorithmic aspects of topology control problems for ad hoc networks. In: Proc. of the 3rd ACM Int. Symposium on Mobile ad hoc Networking and Computing. (2002) 123–134
5. Liu, J., Li, B.: Distributed topology control in wireless sensor networks with asymmetric links. In: Proc. of IEEE Globecom 2003. (2003) 1257–1262
6. Tseng, Y., Chang, Y., Tzeng, B.: Energy-efficient topology control for wireless ad hoc sensor networks. J. of Information Science and Engineering 20 (2004) 27–37
7. Wu, J., Li, H.: On calculating connected dominating set for efficient routing in ad hoc wireless networks. In: Proc. of the 3rd ACM Int. Workshop on Discrete algorithms and methods for mobile computing and communications. (1999) 7–14
8. Hara, T.: Replica allocation methods in ad hoc networks with data update. Mobile Networks and Applications 8 (2003) 343–354
9. Chen, B., Jamieson, K., Balakrishnan, H., Morris, R.: Span: An energy-efficient coordination algorithm for topology maintenance in ad hoc wireless networks. Wireless Networks 8 (2002) 481–494
10. Aspnes, J., Eren, T., Goldenberg, D.K., Morse, A.S., Whiteley, W., Yang, Y.R., Anderson, B.D.O., Belhumeur, P.N.: A theory of network localization. To appear, IEEE Transactions on Mobile Computing (2006)
11. Hohberg, W.: How to find biconnected components in distributed networks. Journal of Parallel and Distributed Computing 9 (1990) 374–386
12. Chaudhuri, P.: An optimal distributed algorithm for computing bridge-connected components. Computer Journal 40 (1997) 200–207
13. Thurimella, R.: Sub-linear algorithms for sparse certificates and biconnected components. Journal of Algorithms 23 (1997) 160–179
14. Swaminathan, B., Goldman, K.: An incremental distributed algorithm for computing biconnected components in dynamic graphs. Algorithmica 22 (1998) 305–329
15. Cidon, I.: Yet another distributed depth-first search algorithm. Inform. Process. Lett. 26 (1988) 301–305
16. Tsin, Y.H.: Some remarks on distributed depth-first search. Inform. Process. Lett. 82 (2002) 173–178
17. Tarjan, R.: Depth first search and linear graph algorithms. SIAM Journal Computing 1 (1972) 146–160
18. ScatterWeb. http://www.scatterweb.net (2006)

Uniquely Localizable Networks with Few Anchors

Zsolt Fekete[1],* and Tibor Jordán[2],**

[1] Computer and Automation Institute, Hungarian Academy of Sciences
fezso@cs.elte.hu
[2] Department of Operations Research, Eötvös University, Pázmány sétány 1/C, 1117
Budapest, Hungary
jordan@cs.elte.hu

Abstract. In the network localization problem the locations of some nodes (called anchors) as well as the distances between some pairs of nodes are known, and the goal is to determine the location of all nodes. The localization problem is said to be solvable (or uniquely localizable) if there is a unique set of locations consistent with the given data. Recent results from graph rigidity theory made it possible to characterize the solvability of the localization problem in two dimensions.

In this paper we address the following related optimization problem: given the set of known distances in the network, make the localization problem solvable by designating a smallest set of anchor nodes. We develop a polynomial-time 3-approximation algorithm for this problem by proving new structural results in graph rigidity and by using tools from matroid theory.

1 Introduction

In the network localization problem the locations of some nodes (called anchors) as well as the distances between some pairs of nodes are known, and the goal is to determine the location of all nodes. This is one of the fundamental algorithmic problems in the theory of wireless sensor networks and it has been in the focus of a number of recent research articles, see e.g. [1,3,16].

The localization problem is said to be solvable (or the network is said to be uniquely localizable) if there is a unique set of locations consistent with the given data. Recent results from graph rigidity theory made it possible to characterize the solvability of the localization problem in two dimensions, assuming that the nodes are in 'general position'. (In what follows we shall also work with this assumption.) In this case the solvability of the problem depends only on the combinatorial properties of the network. In the *graph of the network* vertices

---

* Supported by the Egerváry Research Group of the Hungarian Academy of Sciences, the European MCRTN Adonet, Contract Grant No. 504438, and by the Communication Networks Laboratory, Pázmány Péter sétány 1/A, Budapest, Hungary, H-1117.
** This work was supported by the Mobile Innovation Center, Hungary, and by the Hungarian Scientific Research Fund grant No. T49671, K60802, T37547, and TS49788.

S. Nikoletseas and J.D.P. Rolim (Eds.): ALGOSENSORS 2006, LNCS 4240, pp. 176–183, 2006.

correspond to nodes, and two vertices are connected by an edge if either the corresponding distance is known, or both endvertices are anchors. See Figure 1. As it was observed earlier [3,16], a two-dimensional network in 'general position' is uniquely localizable if and only if it has at least three anchors and the graph of the network is globally rigid (or uniquely realizable).

In this paper we address the following related optimization problem: given the set of known distances in the network, make the localization problem solvable by designating a smallest set of anchor nodes. (This problem was also posed in [16] as an open question.) We develop a polynomial-time 3-approximation algorithm for this problem by proving new structural results in graph rigidity and by using tools from matroid theory. Due to space limitations, we focus on the combinatorial aspects. For more details and other related results see the full version [4] and the list of references.

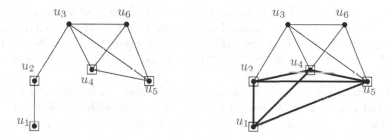

Fig. 1. A uniquely localizable network with six nodes and its graph. In the network edges correspond to known distances between pairs of nodes. The anchor nodes u_1, u_2, u_4, u_5 are indicated by boxes. This is a smallest anchor set which can guarantee solvability for the given set of distances.

2 Rigid and Globally Rigid Graphs

In this section we give a brief summary of the basic definitions and results concerning rigid and globally rigid graphs. See [7,14,17] for more details on rigid graphs and frameworks. For a graph $G = (V, E)$ and a subset $X \subseteq V$ let $E_G(X)$ denote the set, and $i_G(X)$ the number, of edges in $G[X]$, that is, in the subgraph induced by X in G. For $F \subseteq E$ we shall use $i_F(X)$ to denote the number of those edges in F which are induced by X.

Let $G = (V, E)$ be a graph and let $F \subseteq E$. We say that F is *sparse* if

$$i_F(X) \leq 2|X| - 3 \text{ for all } X \subseteq V \text{ with } |X| \geq 2. \tag{1}$$

It is well-known that the sparse subsets of E form the independent sets of a matroid $\mathcal{R}(G)$ on groundset E, with rank function r.

Roughly speaking, a graph is *rigid* if, when realized as a bar-and-joint framework in 'general position', it has no non-trivial continous deformation which preserves the bar lengths. The following fundamental theorem of Laman gives a combinatorial characterization of rigidity in R^2.

Theorem 1. *[9] A graph $G = (V, E)$ is rigid in R^2 if and only if $r(E) = 2|V| - 3$.*

Thus the matroid $\mathcal{R}(G)$ is called the *rigidity matroid* of G. We say that a graph $G = (V, E)$ is *M-independent* if E is independent in $\mathcal{R}(G)$. We have the following formula for the rank of a set of edges of G. A *cover* of $G = (V, E)$ is a collection of subsets $\mathcal{X} = \{X_1, X_2, \ldots, X_t\}$ of V, each of size at least two, such that $\{E_G(X_1), E_G(X_2), \ldots, E_G(X_t)\}$ partitions E. The cover is *non-trivial* if $t \geq 2$. The *value* of the cover is equal to $val(\mathcal{X}) = \sum_{i=1}^{t}(2|X_i| - 3)$.

Theorem 2. *[12] Let $G = (V, E)$ be a graph. The rank of a non-empty set $E' \subseteq E$ of edges in $\mathcal{R}(G)$ is given by*

$$r(E') = \min val(\mathcal{X}),$$

where the minimum is taken over covers $\mathcal{X}$ of (V, E').

Given a graph $G = (V, E)$, a subgraph $H = (W, C)$ is said to be an *M-circuit* in G if C is a circuit (i.e. a minimal dependent set) in $\mathcal{R}(G)$. In particular, G is an *M-circuit* if E is a circuit in $\mathcal{R}(G)$. Using Theorem 1 we may deduce that G is an M-circuit if and only if $|E| = 2|V| - 2$ and $E - e$ is sparse for all $e \in E$. $G - e$ is rigid for all $e \in E$.

Given a matroid $\mathcal{M} = (E, \mathcal{I})$, we define a relation on E by saying that $e, f \in E$ are related if $e = f$ or if there is a circuit C in $\mathcal{M}$ with $e, f \in C$. It is well-known that this is an equivalence relation. The equivalence classes are called the *components* of $\mathcal{M}$. If $\mathcal{M}$ has at least two elements and only one component then $\mathcal{M}$ is said to be *connected*.

We say that a graph $G = (V, E)$ is *M-connected* if its rigidity matroid $\mathcal{R}(G)$ is connected. For example, complete graphs K_m with $m \geq 4$ and complete bipartite graphs $K_{3,m}$ with $m \geq 4$ are M-connected. Note that M-connected graphs are rigid [7]. The *M-components* of G are the subgraphs of G induced by the components of $\mathcal{R}(G)$. For example, the graph in Figure 1 has three M-components: the edges u_1u_2, u_2u_3, and the K_4. Note that the M-components of G are induced subgraphs. For more examples and basic properties of M-connected graphs see [7].

Globally rigid graphs were characterized in [7], see also [5].

Theorem 3. *[7] A graph on at least four vertices is globally rigid in R^2 if and only if it is 3-connected and M-connected.*

3   The M-Connected Relaxation of the Anchor Minimization Problem

The previous discussions and Theorem 3 imply that the *anchor minimization problem* can be reformulated as the following purely combinatorial problem: given a graph $G = (V, E)$, find a smallest set $P \subseteq V$, $|P| \geq 3$, for which $G + K(P)$, the graph obtained from G by adding a complete graph on vertex set P, is 3-connected and M-connected.

To find an approximate solution for the anchor minimization problem we first neglect the 3-connectivity condition and consider its 'M-connected relaxation'. (Note that the complexity status of both problems is still open.)

The following lemma is easy to prove by standard matroid techniques.

Lemma 1. *Let $\mathcal{M} = (E, \mathcal{I})$ be a matroid with components $E_1, E_2, \ldots, E_t$. Then*
(i) $r(\mathcal{M}) = \sum_1^t r(E_i)$, and
(ii) if $r(\mathcal{M}) = \sum_1^q r(F_i)$ for some partition $F_1, F_2, \ldots, F_q$ of E and E_i is a component of $\mathcal{M}$ for some $1 \leq i \leq t$, then $E_i \subseteq F_j$ for some $1 \leq j \leq q$.

The next lemma can be deduced from Theorem 2, Lemma 1, and the fact that the M-components are induced rigid subgraphs.

Lemma 2. *$G = (V, E)$ is M-connected if and only if $val(\mathcal{X}) \geq 2|V| - 2$ for all non-trivial covers $\mathcal{X}$ of G.*

The following key lemma characterizes the feasible solutions of the 'M-connected relaxation'.

Lemma 3. *Let $G = (V, E)$ be a graph, let $\mathcal{H} = \{H_1, H_2, \ldots, H_l\}$ be the M-components of G, and let $P \subseteq V$ with $|P| \geq 4$. Then $G + K(P)$ is M-connected if and only if*

$$2|V| - 2 \leq 2|Z| - 3 + \sum_{H_i \in \mathcal{H}: V(H_i) \cap (V - Z) \neq \emptyset} (2|V(H_i)| - 3) \qquad (2)$$

holds for all $Z \subset V$ with $P \subseteq Z$, $Z \neq V$.

Proof. First suppose that $G + K(P)$ is M-connected. Since $\mathcal{H}$ is a cover of G and $P \subseteq Z$, $\{Z\} \cup \{H_i \in \mathcal{H} : V(H_i) \cap (V - Z) \neq \emptyset\}$ is a cover of $G + K(P)$. This cover is non-trivial, since $Z \neq V$. Thus (2) follows from Lemma 2.

To prove the other direction suppose, for a contradiction, that (2) holds but $G' = G + K(P)$ is not M-connected. Let $\mathcal{H}' = \{H_1', H_2', \ldots, H_q'\}$ denote the M-components of G'. Since complete graphs on at least four vertices are M-connected, and $|P| \geq 4$, it follows that $G'[P]$ is M-connected. Thus there is an M-component H_1', say, for which $P \subseteq V(H_1')$.

Now consider a graph H on vertex set V. We claim that $H = H_j'$ for some M-component H_j' of G' with $2 \leq j \leq q$, if and only if $H = H_i$ for some M-component H_i of G with $V(H_i) \cap (V - V(H_1')) \neq \emptyset$. To see this focus on an M-component $H_j' \in \mathcal{H}'$ with $j \geq 2$. Since $P \subseteq V(H_1')$ and $E_{G'}(H_1') \cap E_{G'}(H_j') = \emptyset$, it follows that $G[V(H_j')]$ is M-connected. Thus, since G' is a supergraph of G and H_1' is an induced subgraph, we must have $H_j' = H_i$ for some $H_i \in \mathcal{H}$ with $V(H_i) \cap (V - V(H_1')) \neq \emptyset$. Now let $H_i \in \mathcal{H}$ with $V(H_i) \cap (V - V(H_1')) \neq \emptyset$. Since H_1' is M-connected in G' and M-components are edge-disjoint, it follows that $V(H_i)$ induces a maximal M-connected subgraph in G'. This proves the claim.

By using the previous claim and Lemma 1(i), and by applying (2) with $Z = V(H_1')$, we obtain

$$2|V|-3 \geq r(G') = 2|V(H'_1)|-3+ \sum_{H_i \in \mathcal{H}: V(H_i) \cap (V-V(H'_1)) \neq \emptyset} (2|V(H_i)|-3) \geq 2|V|-2, \quad (3)$$

a contradiction.   □

Let $\tilde{\mathcal{H}} = (V, \mathcal{E})$ be the hypergraph obtained from $\mathcal{H}$ by replacing each set H_i by $2|V(H_i)| - 3$ copies of $V(H_i)$, $1 \leq i \leq t$. For some $X \subseteq V$ let $e_{\tilde{\mathcal{H}}}(X)$ denote the number of hyperedges $e \in \mathcal{E}$ with $e \cap X \neq \emptyset$. By letting $S = V - Z$ in Lemma 3 and using the above definitions we obtain:

Lemma 4. Let $G = (V, E)$ be a graph, let $\mathcal{H} = \{H_1, H_2, ..., H_t\}$ be the M-components of G, and let $P \subseteq V$ with $|P| \geq 4$. Then $G + K(P)$ is M-connected if and only if

$$e_{\tilde{\mathcal{H}}}(S) \geq 2|S| + 1 \qquad (4)$$

holds for all non-empty subsets $S \subseteq V - P$.

4   The Matroid Matching Problem

Let $\mathcal{M}$ be a matroid on gound-set S and suppose that S is partitioned into a set A of pairs. A subset $M \subseteq A$ is a *matroid matching* if the union of the pairs in M is independent in $\mathcal{M}$. A set $P \subseteq S$ is a *co-matching* if $S - P$ is the union of pairs of a matroid matching. In the *matroid matching problem* the goal is to find a largest matroid matching, see [15, Chapter 43], or equivalently, to find a smallest co-matching. Lovász [11] has shown that this problem may require exponential time in general but can be solved polynomially if the matroid is presented by a set of vectors in some linear space.

We claim that the 'M-connected relaxation', i.e. the problem of finding a smallest set P for which $G+K(P)$ is M-connected, is a special case of the smallest co-matching problem. To see this consider the bipartite graph G^* obtained from the bipartite incidence graph of $\tilde{\mathcal{H}}$ by splitting each vertex $u_i \in V$ into two vertices u'_i, u''_i. Let U denote the color class containing the split vertices. See Figure 2. It is not difficult to see that there is a one-to-one correspondence between the subsets $P \subseteq V$ satisfying (4) and the subsets $P' \subseteq U$ for which $U - P'$ consist of pairs of split vertices and which satisfy the *strong Hall condition* in G^*. (A subset of U is said to satisfy the strong Hall condition if all non-empty subsets of U satisfy the Hall condition with strict inequality.)

We say that a hypergraph $H' = (V, \mathcal{E}')$ is a *hyperforest* if $|\cup \mathcal{F}| \geq |\mathcal{F}| + 1$ for all $\emptyset \neq \mathcal{F} \subseteq \mathcal{E}'$. Lorea [10] proved that in a hypergraph $H = (V, \mathcal{E})$ the edge sets of the hyperforest subhypergraphs of H form a family of independent sets of a matroid on ground set $\mathcal{E}$. This matroid is the *hypergraphic matroid* $\mathcal{M}(H)$ of H. It is easy to see that a subhypergraph of H is a hyperforest if and only if the corresponding set of vertices in the bipartite incidence graph of H satisfies the strong Hall condition.

The above discussion, the construction of G^*, and Lemma 4 imply that the problem of finding a smallest set P for which $G + K(P)$ is M-connected can be

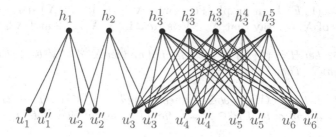

Fig. 2. The bipartite graph G^* obtained from the graph G in Figure 1

formulated as finding a smallest co-matching in a hypergraphic matroid (whose defining hypergraph is obtained from the dual hypergraph of $\tilde{\mathcal{H}}$ by duplicating every hyperedge).

Hypergraphic matroids are known to be linear, but it is not known how to find a suitable linear representation. Thus the complexity status of the matroid matching (and co-matching) problem in hypergraphic matroids is still open. However, as we shall see, the greedy algorithm provides a good approximation.

5   The Approximation Algorithm for the Anchor Minimization Problem

The input of the algorithm is a graph $G = (V, E)$ with at least four vertices. The algorithm has two phases. In each phase we apply a greedy deletion procedure. In the first phase we compute an inclusionwise minimal set $P \subseteq V$ for which $G' = G + K(P)$ is M-connected and $|P| \geq 4$. In the second phase we compute an inclusionwise minimal set $P' \subseteq V$ for which $G' + K(P')$ is 3-connected. The output is the set $P \cup P'$.

Let OPT denote the size of an optimal solution. To see that this is a 3-approximation algorithm for the anchor minimization problem first we show that the first phase is a 2-approximation algorithm for the 'M-connected relaxation', provided that G is not M-connected. This follows from the next lemma and our previous observations.

Lemma 5. *Let $\mathcal{M}$ be a matroid on ground-set S and suppose that S is partitioned into a set A of pairs. Let P' be an inclusionwise minimal co-matching and let P be a smallest co-matching in $\mathcal{M}$ with respect to A. Then $|P'| \leq 2|P|$.*

Proof. Let $X = S - P'$. The minimality of P' implies that $r(X + e) \leq r(X) + 1$ for all pairs $e \in E$ with $e \subseteq P'$. Thus $r(S) \leq |X| + |S - X|/2$, and hence $|P'| = |S - X| \leq 2|S| - 2r(S)$. On the other hand, since $S - P$ is independent, we have $|P| \geq |S| - r(S)$. This gives $|P'| \leq 2|P|$. $\square$

In the second phase we compute a smallest set P' for which $G' + K(P')$ is 3-connected. This follows from the following observations.

Let $H = (V, E)$ be a graph. For some $X \subseteq V$ let $N(X)$ denote the set of neighbours of X. We say that $X \subset V$ is *tight* if $|N(X)| = 2$ and $X \cup N(X) \neq V$.

Lemma 6. *Let $H = (V, E)$ be 2-connected and let $X, Y \subset V$ be distinct minimal tight sets in G. Then $X \cap Y = \emptyset$.*

Lemma 7. *Let $H = (V, E)$ be 2-connected and let $P' \subseteq V$. Then $H + K(P')$ is 3-connected if and only if $P' \cap X \neq \emptyset$ for all minimal tight sets X of H.*

It follows from Lemmas 6 and 7 that an inclusionwise minimal set P' for which $H + K(P')$ is 3-connected is in fact a smallest set.

Theorem 4. *The above algorithm is a 3-approximation algorithm for the anchor minimization problem. Its running time is $O(n^3)$, where n is the number of vertices of G.*

Proof. Clearly, the output $P \cup P'$ is a feasible solution. Since the first phase is a 2-approximation algorithm for a relaxed problem, we have $|P| \leq 2OPT$. The second phase finds a smallest set P' for which $G' + K(P')$ is 3-connected. Note that G' is 2-connected, since it is M-connected. Since G' is a supergraph of G, this gives $|P'| \leq OPT$. Hence $|P \cup P'| \leq |P| + |P'| \leq 2OPT + OPT = 3OPT$.

To bound the running time we need the following observations. The first phase requires at most n M-connectivity tests, which takes $O(n^2)$ each by using the algorithm of [2].

In the second phase we can compute the minimal tight sets in G' by computing the 3-connected components of G' in linear time by using the algorithm of [6]. This gives the bound $O(n^3)$ in total.   □

6   Concluding Remarks

In this paper we considered the anchor minimization problem: given a graph $G = (V, E)$, find a smallest set $P \subseteq V, |P| \geq 3$, for which $G + K(P)$ is 3-connected and M-connected. We gave a polynomial-time 3-approximation algorithm for this problem.

As we noted earlier, the complexity status of this problem as well as its 'M-connected relaxation' remains open. We note that in a recent paper Makai [13] gave a good characterization (a minimax formula) for the maximum size of a matroid matching in a hypergraphic matroid. This indicates that the 'M-connected relaxation' might turn out to be polynomially solvable. Furthermore, in the full version of this paper [4], we give a randomized polynomial-time algorithm which can optimally solve this relaxation and leads to a randomized 2-approximation algorithm for the anchor minimization problem.

Further open problems include the weighted cases of these minimization problems and the versions where the goal is to enlarge a given set of anchors to ensure solvability, or to make some specified nodes of the network uniquely localizable (see [8]).

References

1. J. ASPNES, T. EREN, D.K. GOLDENBERG, A.S. MORSE, W. WHITELEY, Y.R. YANG, B.D.O. ANDERSON, P.N. BELHUMEUR, A theory of network localization, *Trans. Mobile Computing*, to appear.

2. A.R. BERG AND T. JORDÁN, Algorithms for graph rigidity and scene analysis, Proc. 11th Annual European Symposium on Algorithms (ESA) 2003, (G. Di Battista, U. Zwick, eds) Springer Lecture Notes in Computer Science 2832, pp. 78-89, 2003.

3. T. EREN, D. GOLDENBERG, W. WHITELEY, Y. R. YANG, A. S. MORSE, B. D. O. ANDERSON, AND P. N. BELHUMEUR, Rigidity, Computation, and Randomization in Network Localization, Proc. of the IEEE INFOCOM Conference, Hong-Kong, March 2004, pp. 2673-2684.

4. Z. FEKETE AND T. JORDÁN, Uniquely localizable networks with few anchors, EGRES TR-2006-07, www.egres.hu/egres/

5. B. HENDRICKSON, Conditions for unique graph realizations, *SIAM J. Comput.* **21** (1992), no. 1, 65-84.

6. J.E. HOPCROFT AND R.E. TARJAN, Dividing a graph into triconnected components, *SIAM J. Comput.* 2 (1973), 135-158.

7. B. JACKSON AND T. JORDÁN, Connected rigidity matroids and unique realizations of graphs, *J. Combinatorial Theory, Ser. B.*, Vol. 94, 1-29, 2005.

8. B. JACKSON, T. JORDÁN, AND Z. SZABADKA, Globally linked pairs of vertices in equivalent realizations of graphs, *Discrete and Computational Geometry*, Vol. 35, 493-512, 2006.

9. G. LAMAN, On graphs and rigidity of plane skeletal structures, *J. Engineering Math.* 4 (1970), 331-340.

10. M. LOREA, Hypergraphes et matroides, *Cahiers Centre Etud. Rech. Oper.* 17 (1975) pp. 289-291.

11. L. LOVÁSZ, The matroid matching problem, in: *Algebraic methods in graph theory*, Proc. Conf. Szeged (1978).

12. L. LOVÁSZ AND Y. YEMINI, On generic rigidity in the plane, *SIAM J. Algebraic Discrete Methods* 3 (1982), no. 1, 91–98.

13. M. MAKAI, Matroid matching with Dilworth truncation, Proc. EuroComb 2005, Berlin.

14. A. RECSKI, *Matroid theory and its applications in electric network theory and in statics*, Akadémiai Kiadó, Budapest, 1989.

15. A. SCHRIJVER, *Combinatorial Optimization*, Springer, Berlin, 2003.

16. A. MAN-CHO SO AND Y. YE, Theory of semidefinite programming for sensor network localization, Proceedings of the Sixteenth Annual ACM-SIAM Symposium on Discrete Algorithms (SODA) 2005.

17. W. WHITELEY, Some matroids from discrete applied geometry, in Matroid theory (J.E. Bonin, J.G. Oxley and B. Servatius eds., Seattle, WA, 1995), Contemp. Math., 197, Amer. Math. Soc., Providence, RI, 1996, 171–311.

A Locating Method for Ubiquitous Robots Based on Wireless Sensor Networks[*]

Namkoo Ha, Minsu Kim, Byeongjik Lee, and Kijun Han[**]

Department of Computer Engineering,
Kyungpook National University, Daegu, Korea
Phone number: 82-53-950-5557
Fax number: 82-53-957-4846
{adama2,kiunsen,leric}@netopia.knu.ac.kr,
kjhan@knu.ac.kr

Abstract. Applications of Ubiquitous Robot (UR) based on wireless sensor networks usually require the location information of all sensor nodes deployed on the sensor field. For this, every sensor node should be aware of its own geographical location, which forces us to pay an expensive cost and make the size of sensor node larger. In this paper, we propose a cheap solution for positioning all sensor nodes without necessitating all sensor nodes in the field equipped with GPS modules. In our method, only some sensor nodes equipped with GPS modules are initially deployed, and the other nodes without GPS modules will find out their locations through communications to GPS nodes. To show effectiveness of our method, we carried out a computer simulation, and observed that all nodes could successfully recognize their locations, which can considerably save the price for implementing wireless sensor networks.

1 Introduction

In most recent years, motivated by the emergence of ubiquitous computing technology as the next generation of computing paradigm, a new class of networked robot - ubiquitous robot (UR) - has been introduced. The URC (Ubiquitous Robot Companion) is the conceptual vision of UR which provides me with the services I need, anytime and anyplace in ubiquitous computing environment [1]. Since it is inherently based on ubiquitous environment with networked sensors and actuators, it can be considered as one of the most important emerging applications of sensor network. Fig. 1 depicts the concept of URC with sensing, processing and acting abilities in wireless sensor networks to overcome the technical constraints and producing costs by utilizing sensors and remote computing server.

By using the network with external cheap sensors embedded in the environment replacing multiple robots equipped sensors, the context-awareness of the robot would be dramatically improved and lessen the burden of hardware cost [2]. Besides, remote

---

[*] This work was supported by grant No. (R01-2005-000-10722-0) from the Basic Research Program of the Korea Science & Engineering Foundation.
[**] Correspondent author.

S. Nikoletseas and J.D.P. Rolim (Eds.): ALGOSENSORS 2006, LNCS 4240, pp. 184–191, 2006.
© Springer-Verlag Berlin Heidelberg 2006

computing server can be used as an external memory and processor of URC, which improves the robot intelligence and expands its applications and services. Researchers are looking into merging robotics and sensor network technologies.

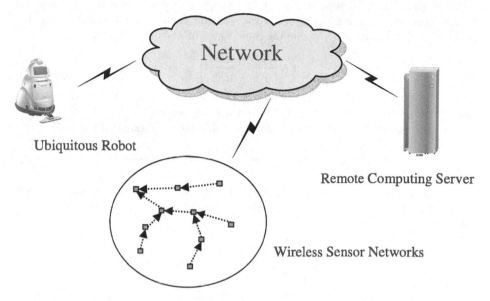

Fig. 1. The concept of Ubiquitous Robot Companion (URC)

In many applications, the UR should make use of the location information of each sensor node distributed on the sensor network. Most works for locating applications are based on the assumption that each sensor node knows its own geographical location [3–5]. To recognize sensor nodes' geographical location, they should be equipped with GPS modules in themselves, which causes an extra cost as well as the higher energy consumption and the larger size [6][7].

To reduce the cost, price, and size of sensor nodes, Dragos Niculescu proposed APS (Ad hoc Positioning System) which assumed that only three landmark nodes know their own locations in the entire field [8]. The nodes which do not know their locations estimate their locations based on the number of hops and relative distance to three landmark nodes, respectively. To estimate this, these landmark nodes flood information about their geographical locations. Although this method can reduce price and size of the nodes, it takes too long for all nodes to recognize their locations. Also, if there is initially even a small error in the location of landmark node, the error becomes accumulated every flooding time, and can finally exceed an allowable limit.

We propose a cost effective locating method, in which the minimal sensor nodes of the deployed nodes in sensing field are equipped with the GPS and the other nodes try to recognize their locations through communications to the nodes with GPS modules.

The organization of this paper is as follows. Section 2 describes our method, and section 3 presents the validation of our method through computer simulations. Finally, section 4 concludes this paper.

2 Proposed Method

Any node can find out its location using location information collected from three or more neighbor nodes that have already known their locations in a two-dimensional field. When N nodes are deployed in the sensing field in a random fashion, the probability that a node exists within the radio range of node n_i is given by

$$p = \frac{S_{n_i}}{S} = \frac{\pi r^2}{M^2} \tag{1}$$

where S is the size of the whole sensing field with a dimension of $M \times M$, S_{ni} is the radio range of node n_i, and r is the radius of the radio range.

The average number of one-hop neighbors within the radio range of node n_i is

$$\overline{D} = \sum_{k=1}^{N-1} k \cdot \binom{N-1}{k} p^k (1-p)^{N-1-k} = (N-1)\frac{\pi r^2}{S} \tag{2}$$

The minimal number of nodes that should be equipped with GPS modules to make it possible for all nodes in the sensing field to know their locations, denoted by N_{Ge}, is given by

$$N_{Ge} = \frac{3S}{\pi r^2} + 1 \tag{3}$$

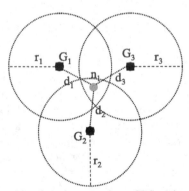

(a) when there are three GPS nodes

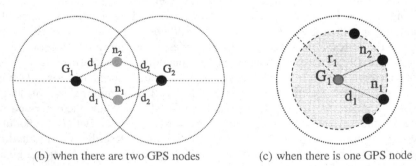

(b) when there are two GPS nodes   (c) when there is one GPS node

Fig. 2. Three scenarios for distribution of GPS nodes

Any node which does not know its own location tries to find out its location using location information collected from three or more neighbor nodes that have already known their locations. Assume that we are given a node without GPS, denoted by $n_i(x,y)$, whose location is not yet known. If we assume that there are three nodes equipped with GPS module around the node $n_i(x,y)$ as shown in Fig. 2(a), denoted by $G_1(x_1, y_1)$, $G_2(x_2, y_2)$, and $G_3(x_3, y_3)$, then the coordinate of node $n_i(x, y)$ is computed by

$$x = \frac{(y_2 - y_3)(d_1^2 - x_1^2 - y_1^2) + (y_3 - y_1)(d_2^2 - x_2^2 - y_2^2) + (y_1 - y_2)(d_3^2 - x_3^2 - y_3^2)}{2\{(y_2 - y_1)(x_3 - x_1) - (y_3 - y_1)(x_2 - x_1)\}} \tag{4a}$$

$$y = \frac{(x_3 - x_2)(d_1^2 - x_1^2 - y_1^2) + (x_1 - x_3)(d_2^2 - x_2^2 - y_2^2) + (x_2 - x_1)(d_3^2 - x_3^2 - y_3^2)}{2\{(y_2 - y_1)(x_3 - x_1) - (y_3 - y_1)(x_2 - x_1)\}} \tag{4b}$$

where d_1, d_2 and d_3 mean the Euclidean distances from $n_i(x,y)$ to the three nodes, respectively, which are given by

$$d_i^2 = (x - x_i)^2 + (y - y_i)^2 \qquad i = 1,2,3 \tag{5}$$

When there are two or less nodes equipped with GPS module around $n_i(x,y)$, its location cannot be directly determined as illustrated in Fig. 2(b) and (c). In this case, its location is iteratively determined by the following neighbor discovery procedure.

REPEAT
1. Each GPS node broadcasts LI(Location Information) to its neighbors within the radio range
2. When a node receives LI,
3. If (GPS node)
   a. Discard received LI
4. Else if ((no-GPS node) & (number of received LI's) 3)
   a. Compute d_i and its location (n_i(x,y)) by $Eq.$ (4) & (5)
   Go to Step 1
UNTIL all nodes are aware of their locations

Fig. 3. Neighbor discovery procedure

As mentioned so far, in our method, only some nodes in the sensing field are equipped with the GPS and the others can recognize their locations through communications to nodes with GPS.

3   Simulations

To validate our method, we carried out simulation study. Fig. 4 shows the simulation model. We randomly deploy 100 nodes in a square space (50 × 50) or 400 nodes in a square space (100 × 100). Table 1 shows some parameters and their values used for simulation. The number of GPS nodes is calculated by Eq. (3).

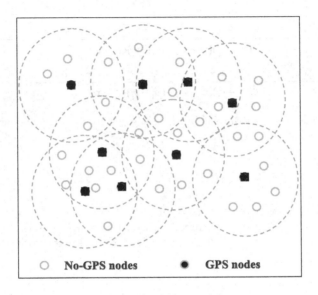

Fig. 4. Simulation model

Table 1. Simulation parameters and their values in four scenarios

Simulation parameters	Case 1	Case 2	Case 3	Case 4
Map size	50 x 50	100 x 100	50 x 50	100 x 100
Radio range	10	10	15	15
The number of total nodes	100	400	100	400
The number of GPS nodes	24	96	11	43
The number of no-GPS node:	76	304	89	357

Fig. 5 shows the average number of neighbor nodes within the radio range of node n_i form Case 1 to 4. A sound agreement is observed between the numerical results obtained by Eq. (2) and the result of computer simulation. The number of neighbor nodes is about 12 and 28 in case 1 and 3, respectively. This means that if we deploy only three GPS nodes among 12 and 28 nodes, in case 1 and 3, respectively, all nodes can identify their locations in several rounds, which can greatly save price and make nodes smaller.

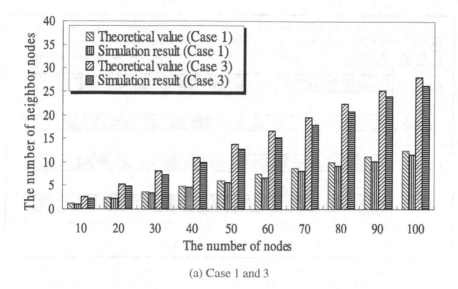

(a) Case 1 and 3

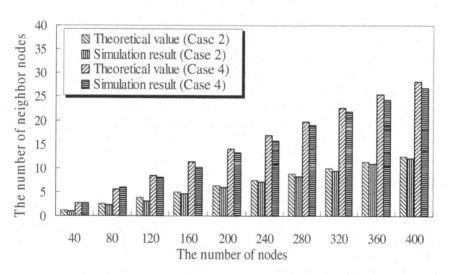

(b) Case 2 and 4

Fig. 5. The number of neighbor nodes

Fig. 6 shows how long it will take for all nodes to be aware of their locations. In this figure, we see that only 52% and 44% of no-GPS nodes can compute their locations in the first round in case 1 and 3, respectively. This figure shows that all no-GPS nodes can recognize their locations in the third round for all four cases.

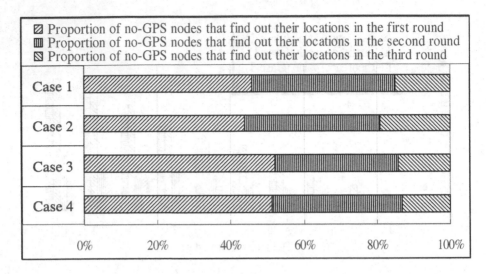

Fig. 6. Proportion of no-GPS nodes that are aware of their locations in each round

4 Conclusions

In this paper, we proposed a considerably cheap solution for locating method in wireless sensor networks. In our method, only a few nodes equipped with GPS modules are deployed in the sensing field, and the other nodes are able to recognize their locations through communications to GPS nodes. Simulation results assure our belief that our method can be successfully applicable to UR based on wireless sensor networks.

References

1. Young-Guk Ha; Joo-Chan Sohn; Young-Jo Cho, "Service-oriented integration of networked robots with ubiquitous sensors and devices using the semantic Web services technology," *Intelligent Robots and Systems, 2005. (IROS 2005). Internatinal Conference on* 2-6 Aug. 2005 Page(s):3947 - 3952
2. H.Kim, Y.-J.Cho and S.-R.Oh, "CAMUS: A middleware supporting context-aware services foe network-based robots," *In Proceeding of the IEEE Workshop on Advanced Robotics and its Social Impacts(ARSO)*, 2005.
3. Deepak Ganesan, Alberto Cerpa, Yan Yu, Deborah Estrin, Wei Ye, Jerry Zhao, "Networking Issues in Wireless Sensor Networks," *Journal of Parallel and Distributed Computing*, Vol. 64, Issue 7, pp. 799 – 814, July 2004.
4. C.Y. Chong and S.P. Kumar, "Sensor networks: evolution, opportunities and challenges", *Proceedings of the IEEE*, Vol. 91, No. 8, August 2003.
5. I. F. Akyildiz, W. Su, Y. Sankarasubramaniam, E. Cayirci, "A survey on sensor networks," *IEEE Communications Magazine*, Vol. 40(8), pp. 102-114, August 2002.

6. L. Doherty, K. J. Pister, L. El Ghaoui, "Convex Position Estimation in Wireless Sensor networks," *Proceedings of IEEE INFOCOM*, Vol. 3, pp. 1655-1663, April 2001.
7. P. Bahl and V. N. Padmanabhan, "RADAR: An in-building RF based user location and tracking system," *Proceedings of IEEE INFOCOM,* pp. 775-784, Tel Aviv, Israel, March, 2000.
8. D. Niculescu and B. Nath, "Ad hoc positioning system (APS)," *Proceedings of IEEE GLOBECOM,* pp. 2926-2931, San Antonio, TX, November 2001.

Declarative Resource Naming for Macroprogramming Wireless Networks of Embedded Systems

Chalermek Intanagonwiwat[1], Rajesh Gupta[2], and Amin Vahdat[2]

[1] Department of Computer Engineering, Chulalongkorn University, Thailand
[2] Department of Computer Science and Engineering,
University of California at San Diego, USA
intanago@cp.eng.chula.ac.th, {rgupta,vahdat}@cs.ucsd.edu

Abstract. Programming Wireless Networks of Embedded Systems (WNES) is notoriously difficult and tedious. To simplify WNES programming, we propose *Declarative Resource Naming* (DRN) to program WNES as a whole (*i.e., macroprogramming*) instead of several networked entities. DRN allows for a set of resources to be declaratively described by their run-time properties, and for this set to be mapped to a variable. Using DRN, resource access is simplified to only variable access that is completely network-transparent. DRN provides both sequential and parallel accesses to the desired set. Parallel, or group, access reduces the total access time and energy consumption because it enables in-network processing. Additionally, we can associate each set with tuning parameters (*e.g.,* timeout, energy budget) to bound access time or to tune resource consumption.

Keywords: Macroprogramming, Naming, Wireless Networks, Embedded Systems, and Sensor Networks.

1 Introduction

WNES consists of a massive number of resource-constraint wireless nodes that are deployed in dynamic, hostile environments. Unlike traditional networks, WNES is property-centric. As such, nodes of interest in WNES are defined by their properties at run-time rather than by their node ids. These characteristics pose two major research challenges to the design of WNES programming.

1. How do we easily and efficiently write the WNES applications?
2. How do we reprogram the network of unattended nodes after deployment?

In this paper, we will focus on the first question of characterizing WNES applications. To simplify WNES programming, we propose *Declarative Resource Naming* (DRN) to program WNES as a whole (*i.e., macroprogramming*) instead of several networked entities. DRN allows programmers to declaratively describe a set of desired resources (nodes) by their run-time properties and to map this set to a variable. To access the desired resources, we can simply refer to the

S. Nikoletseas and J.D.P. Rolim (Eds.): ALGOSENSORS 2006, LNCS 4240, pp. 192–199, 2006.

mapped variable. Therefore, remote resource access is simplified to only variable access that is completely network-transparent. DRN provides both sequential and parallel access to the desired set. Parallel access reduces the total access time and energy consumption because it enables data aggregation in the network. Additionally, we can associate each set with tuning parameters (*e.g.,* timeout, energy budget) to bound access time or to tune resource consumption.

Given that WNES may be deployed in dynamic, hostile environments, and also that we may not be able to physically reach the nodes, it is necessary that we can remotely program these unattended nodes. Systems based on code migration are preferable because programs can be propagated to target nodes without human intervention. Examples of such systems include Smart Messages (SM) [3], and Mate [6]. Even though reprogramming the network is not our focus in this paper, we have completed our preliminary implementation of a DRN run-time library using Smart Messages (SM) that runs on iPAQs equipped with 802.11 radios. SM is an appropriate choice because it supports program migration, a necessary capability for reprogramming the network. Undoubtedly, there are other reprogrammable platforms such as Mate. However, we select SM because the library can be implemented in Java, a well known language.

2   What Is the Right Abstraction?

Traditionally, there are two programming styles in computer literature: declarative and imperative. Declarative programming fully abstracts out all algorithmic details. Programmers only specify what they want rather than how to algorithmically obtain the results. The translator and optimizer will then fill in the algorithms. Automatic generation of algorithmic details can be efficient for simple and specific tasks (*e.g.,* database), but is questionable for others. Examples of such an SQL-based approach include COUGAR [1]. Despite its simplicity, declarative programming is not applicable for every WNES application. Imperative programming is more appropriate for complex tasks where efficient algorithmic details are either not obvious, or not easy to generate automatically. For example, it is difficult or even impossible to implement Kalman filters or maximum likelihood algorithms for estimating object locations in SQL because SQL is not designed for expressing algorithmic details.

Declarative and imperative programming function well within their domain and complement one another. Integration of declarative constraints and imperative constructs can form a powerful programming paradigm suitable for both domains. In this paper, we propose that such integration is possible if the declarative abstraction is applied only to some parts of the program.

In general, potential targets for abstraction are: 1) parts that are unrelated to the core algorithms; 2) common to applications, and; 3) tedious for programmers. To identify the abstractable parts, a basic understanding of WNES programs is required. Typically, programs are collections of operations on variables and resources. Given that variables are more frequently accessed, programming languages provide a simpler way to access variables than to access resources.

Not surprisingly, traditional resource access is more tedious, especially in networked systems where there exists a distinction between local and remote resources. Resources are normally bound to nodes that are known a priori. Therefore, in order to specify the remote resources that are of interest, node ids are required. If the node ids are not known, resource discovery is needed. As a result, programmers are required to work on several programming details (*e.g.,* networking, resource discovering, resource accessing).

WNES programming is still more labor-intensive because the resources of interest are specified by their properties at run-time rather than node ids. For example, we may want to access sensors on a particular hill only when the temperature is more than 30 degrees Celsius. In this case, resource discovery in WNES becomes necessary and common rather than optional. The resource property is highly dynamic because the environment – where the temperature can drop below 30 degrees Celsius at any moment – is hostile and volatile. Some resource bindings or mappings may have to be invalidated because the bound resources may no longer match the desired property. But even if the resource property does not change, bound resources may not be accessible because of network dynamics such as node mobility. WNES programs are required to handle changes, invalidate bindings, discover equivalent resources, and bind the newly discovered resources. Given that the above events are frequent in WNES, these resource handlings (*e.g.,* discovering, accessing, rebinding, and networking) are tedious to programmers. Therefore, the resource-related parts of the WNES program are reasonable choices for our declarative abstraction.

3   Declarative Resource Naming

To simplify the programming tasks for WNES, we propose a scheme that will program the WNES as a unit. Particularly, we consider WNES a single abstract machine. Although physically scattered, all resources are on the same machine in our model. Given this single machine model, there is no notion of networking, being remote, or local.

3.1   Resource Variable

WNES programming can be simplified by making a resource access as simple as a variable access. In order to do this, we propose *resource variables* (*i.e.,* variables that are mapped and referred to actual nodes). For example, one can write a program to read a light sensor and to control a camera as follows.

```
Resource R, X;
printf("light intensity=%f", R->light);
X->camera=off;
```

In the above example, we assume that the resource variable R contains a light sensor and the resource variable X contains a camera. To read the light intensity, we can simply refer to $R->light$. Similarly, the camera can be turned off by assigning off to $X->camera$. There is no need for algorithmic detail of resource controls and operations. This example shows that our approach is not only for retrieving data and for pushing data to desired nodes but also for controlling them.

3.2 Declarative Constraint

Understandably, one may wonder to which physical nodes (or resources) these variables (R and X) are precisely bound and how programmers know about the individual sensor types. Rather than specify the node ids for binding, a target resource's desired property can be declaratively indicated with a boolean expression or a predicate. For example, we can specify that R will be bound to light-sensor nodes within the forest with temperatures greater than 30 degrees Celsius.

```
Resource R = <within(forest) && temperature > 30 && exist(light)>
Resource X = <a(b,c)!=0 && exist(camera)>
```

Given that more than one nodes can match a specified expression, a resource variable is referred to as a set of matching nodes rather than a single one. Location and temperature are local properties (of a node) that are used to determine the node's membership in the set R. Furthermore, we also allow user-defined boolean functions (e.g., function $a()$) in our expression. Such a flexible expression is generally powerful and sufficient for various complex conditions.

3.3 Resource Access

In this section, we illustrate the need for various types of DRN resource access that can be used in different situations. Their advantages and disadvantages are also provided as a guideline for selecting the resource access type that is most suitable for a particular task. We propose two approaches for accessing multiple matching nodes: *sequential* and *parallel*.

- **Sequential Access.** Each element in a set can be referred to using an iterator (similar to an iterator in C++ standard template library). The iterator enables sequential and selective access of resources. For example, one can sequentially read the light intensity of each resource in the set R as follows.
  ```
  Resource R;
  Iterator i;
  foreach i in R {printf("light intensity = %f\n", i->light);}
  ```
  However, the sequential readings cannot represent a snapshot of the desired target because the delay in accessing the whole set sequentially can be significant. In particular, the total delay is essentially the summation of all individual access time. Nevertheless, this individual approach is still useful, especially when only some elements in the set are accessed.
- **Parallel Access.** Conversely, in this approach, all resources in the set are simultaneously accessed. This parallel access can be specified using a direct reference to the resource variable as follows.

```
Resource R;
printf("light intensity=%f", R->light);
```

In the above example, the program prints out the light intensity of all nodes in R. The total delay using this parallel approach is reduced to the longest delay of an access. The parallel approach not only reduces the total access time but also provides a much better snapshot of the desired target. Additionally, unlike the sequential approach, this parallel approach exposes an opportunity for the underlying system to perform in-network processing (*e.g.*, data aggregation) that can significantly reduce a system's overall energy consumption [4]. An example of data aggregation functions is $max(A)$ whereby the maximum element in A is returned.

```
Resource R;
printf("max light intensity = %f", max(R->light));
```

Ideally, the system expends energy only on delivering that max element, not on the others. This delivery can be practically approximated by in-network suppression of the elements whose values are less than that of the previously seen elements of the same access. Suppression will be ineffective or even impossible if the resources are accessed in sequence rather than in parallel.

3.4   Resource Binding

Our model supports two binding types: *dynamic* and *static*.

– **Dynamic Binding.** In our paradigm, code does not need to be written to maintain binding between the physical resources and resource variables. Given that the resource property is constantly changing, rebinding the set of matching nodes is laborious. For example, the set of resources R at time t_1 can be completely different from the set of resources R at time t_2.

  Rather, it is desirable to simply provide the declarative expression that is associated with the resource variable to describe the resources of interest. In general, a reference to a resource variable implies a resource access. Our semantic of a resource-variable access is rather strict in a sense that the access is only performed on the resource that matches the declarative expression at the time of access. Furthermore, changes in the set of matching nodes do not require attention from programmers. As a result, to conform with this strict semantic, the underlying system may need to expend significant overhead and excessive energy consumption for ensuring that this reactive binding is up to date. Therefore, we propose options or *tuning knobs* for lessening the semantic in order to save energy. For example, programmers can lessen the semantic by allowing access if the resource is bound in the last t seconds.

```
Resource R = <expression, last_bound_time > now-t>
```

Furthermore, programmers can even specify an energy budget to bound the energy consumption of a resource access.

```
Resource R = <expression, energy_budget = 100>
```

Other tuning knobs are currently under investigation.

– **Static Binding.** Although the above dynamic binding of resources seems reasonable, one may notice that there are situations where dynamic bindings may not be appropriate. Specifically, we may want to access the previously matched resources that are no longer matched. For example, we may have turned on cameras in area A. However, after a period of time, we may want to turn them off, but some cameras have since been moved out of the area. If area A is included in our declarative expression, those cameras that have since been transferred will no longer match the expression. As a result, we may be unable to turn off the relocated cameras directly using the resource variable.

Our solution is to provide explicit instructions for memorizing matching nodes. We propose two explicit mechanisms: the *static* resource and the *iterator*.

Using the static resource variable, we can specify which resources are statically bound. The static resource variable will not be rebound in any circumstances. Therefore, we can maintain any set of resources even though they are no longer matched to the expression.

```
Resource R1;
Static Resource R2=R1; /* R1 changes over time but R2 does not*/
```

Furthermore, the static resource is intended for memorizing the entire set of matching nodes. To memorize only one resource, an iterator is more appropriate. The value of an iterator does not automatically change without an explicit assignment.

3.5   Access Timeout

Regardless of binding type, there is no guarantee that every WNES resource access will succeed. Unfortunately, WNES resource access time is unbound, and access failures are usually unavoidable because of network dynamics. Given that there is no response after unbound access time and failures, they cannot be easily differentiated. [1] Timeout is usually a common technique for handling such problems. Therefore, we propose associating a resource variable with an access timeout. In this model, the access time is monitored for each access. Once an access has timed out, an exception is raised (similar to Java exceptions). It is necessary that the method for handling a time-out is explicitly specified in the *catch* statement.

```
Resource R = <expression1, timeout = 10>
Iterator i = R->first_element;
try { printf("light intensity = %f", i->light);}
catch(TimeoutException) {printf("can't access the light sensor");}
```

4   Related Work

WNES programming has begun to receive attention during the last few years. However, our work has been informed and influenced by a variety of other research efforts, which we now describe.

---

[1] This problem is similar to that of TCP. Packet loss and unbound acknowledgment delay are handled using timeout.

Our work is mostly influenced by Spatial Programming (SP) [2] and Spatial View (SV) [7]. However, SP supports only sequential resource access, whereas DRN supports both sequential and parallel access. Additionally, SP is purely imperative programming, but DRN is a hybrid between declarative and imperative programming.

The recent SV version [7] supports parallel access via compiler optimization. As a result, data aggregation in SV is possible but implicit, not explicit as in DRN. Furthermore, DRN programmers can write a new in-network processing function by deriving from Aggregation class (see [5] for more details) but it is not obvious how a new in-network processing function can be written in SV. While SV node naming is based on space and sensor-node type, DRN node naming is based on declarative predicate. Furthermore, DRN provides tuning knobs for balancing the desire to abstract network details while giving programmers control over resource usage but SV does not.

The abstract region [8] work focuses on a wider definition of space. Specifically, space in the abstract regions can be physical or logical. For example, the logical space can be defined by the number of hops in communication. This example indicates that, unlike our work, the abstract region does not intend to hide the networking details from programmers. Furthermore, the space is simply an applicable attribute (albeit a very useful one) for our declarative description of resources. Therefore, the space is hardly considered the focus of our work.

Programming WNES as a unit has also been explored earlier by several research efforts, including COUGAR [1]. While the above efforts propose programming WNES as a database using SQL, we propose programming WNES as a single abstract machine. Furthermore, SQL is designed for expressing the desired data, but not for expressing the algorithmic details. Conversely, DRN is designed for both purposes.

5   Conclusions

We believe that, to efficiently develop WNES applications, appropriate programming abstractions are neccessary. DRN is one such abstraction that integrates declarative constraints with imperative constructs to form a powerful programming paradigm suitable for macroprogramming WNES. We have completed our preliminary implementation of a DRN run-time library using Smart Messages (SM) that can run on iPAQs communicating with 802.11 radios. Due to a page limit, our implementation and evaluation are not included in this paper. Please see [5] for more details.

Furthermore, given network transparency, our abstraction is independent of the underlying platforms. Therefore, our approach is applicable for macroprogramming other wired or wireless networks as well. However, this network transparency feature of DRN implies that DRN is not for low-level programming or implementing a protocol that requires a distinct notion between being remote and local.

In the future, we intend to further explore the design space of DRN such as other tuning knobs. Additionally, we plan to implement other applications using DRN and to conduct more extensive evaluation in order to better realize DRN's full potential.

References

1. Philippe Bonnet, Johannes Gehrke, Tobias Mayr, and Praveen Seshadri. Query processing in a device database system. Technical Report TR99-1775, Cornell University, October 1999.
2. Cristian Borcea, Chalermek Intanagonwiwat, Porlin Kang, Ulrich Kremer, and Liviu Iftode. Spatial programming using smart messages: Design and implementation. In *Proceedings of the 24th International Conference on Distributed Computing Systems (ICDCS 2004)*, Tokyo, Japan, March 2004.
3. Cristian Borcea, Deepa Iyer, Porlin Kang, Akhilesh Saxena, and Liviu Iftode. Cooperative Computing for Distributed Embedded Systems. In *Proceedings of the 22nd International Conference on Distributed Computing Systems (ICDCS)*, pages 227–236, July 2002.
4. Chalermek Intanagonwiwat, Ramesh Govindan, and Deborah Estrin. Directed diffusion: A scalable and robust communication paradigm for sensor networks. In *Proceedings of the Sixth Annual ACM/IEEE International Conference on Mobile Computing and Networking (Mobicom'2000)*, Boston, Massachusetts, August 2000.
5. Chalermek Intanagonwiwat, Rajesh Gupta, and Amin Vahdat. Declarative resource naming for macroprogramming wireless networks of embedded systems. Technical Report CS2005-0827, University of California at San Diego, May 2005.
6. P. Levis and D. Culler. A tiny virtual machine for sensor networks. In *Proceedings of the ACM Conference on Architectural Support for Programming Languages and Operating Systems (APLOS)*, October 2002.
7. Yang Ni, Ulrich Kremer, Adrian Stere, and Liviu Iftode. Programming ad-hoc networks of mobile and resource-constrained devices. In *Proceedings of the ACM PLDI*, Chicago, Illinois, USA, June 2005.
8. Matt Welsh and Geoff Mainland. Programming sensor networks using abstract regions. In *Proceedings of the First USENIX/ACM Symposium on Networked Systems Design and Implementation (NSDI 2004)*, March 2004.

Equalizing Sensor Energy and Maximising Sensor Network Lifespan Using RETT*

Khaled Matrouk[1] and Bjorn Landfeldt[2]

[1] School of Electrical and Information Engineering
The University of Sydney,
Sydney 2006, Australia and
National ICT Australia
matrouk@ee.usyd.edu.au
[2] School of Information Technologies
The University of Sydney,
Sydney 2006, Australia and
National ICT Australia
bjornl@it.usyd.edu.au

Abstract. Lately, sensor networks have received much attention in the research community. One of the aims of these networks is to make them self-configuring and self-healing so that they can be easily deployed and self-sufficient. A fundamental limitation with sensor networks is their limited life span because of their battery operation. To that end, much research has been carried out to maximize the life span of sensor networks through energy conservation. In this paper; we investigate energy conservation from a different angle. We introduce RETT, a protocol that aims at equalising energy in such a way that the entire network remains operational at maximal time in case the application would so require. When the entire network is operational, it is certain that events from any sensor in the network will find a delivery path through the network and that actuation of actuators in response to sensed events will be carried out. We focus on the algorithms that enable this and provide experimental results that show how they save on energy compared with the well-known LEACH algorithm.

Keywords: Energy efficiency, Clustering method, Lifetime; Wireless sensor network.

1 Introduction

As a result of recent advances in wireless communication and sensor technologies, wireless sensor networks have become a popular area of research [1]. In wireless sensor networks, the sensors are often intended to work in inaccessible or hostile environments.

Thus, it is difficult to recharge or replace the battery power of the sensors, at least on a regular basis [2]. Therefore, conserving energy and prolonging system lifetime is one of the major challenges of sensor networks.

---

* National ICT Australia is funded through the Australian Government's *Backing Australia's Ability* initiative, in part through the Australian Research Council.

S. Nikoletseas and J.D.P. Rolim (Eds.): ALGOSENSORS 2006, LNCS 4240, pp. 200–207, 2006.
© Springer-Verlag Berlin Heidelberg 2006

Besides maximizing the lifetime of the sensor nodes, it is advantageous to equalize or distribute the energy dissipated throughout the entire network in order to minimize maintenance times and maximize overall system performance. Furthermore, equalizing the energy consumption will keep approximate network wide energy equivalence by replacing heavily dissipated nodes with their unused or less used neighbours.

Over the past few years there have been numerous proposals of power efficient routing protocols for sensor networks i.e. [3-8]. It has been shown that techniques such as hierarchical formation, clustering and intelligent power aware routing can significantly extend the life span of the network. In this paper we focus on maximizing the life span of the entire network, rather than individual nodes. We believe that it will be crucial for many applications that there are no sections of the network that have lost connectivity due to critical nodes running out of power. To achieve this goal, we use the advantages of clustering and self-organizing methods coupled with routing mechanisms that strive to achieve equalization of the available power throughout the network.

Routing based on Energy – Temperature Transformation, **RETT**, is our energy - equalizing routing algorithm designed for wireless sensor networks. In RETT, we transform the expected lifetime of each node to an equivalent temperature, and then by using the heat conduction formulas, we find the hottest path for sending data to the sink, which will not always be the shortest path. We have assumed a virtual grid area in two dimensions, where each square inside that area has a group of sensor nodes forming a cluster and each cluster has a selected head that deals with sending, receiving, and routing the data in a power equalizing manner. As the lifetime of a cluster is defined as the time required for all nodes in the cluster to lose all of their energy, the cluster head with the greatest residual energy density or temperature relays packets on behalf of other neighbour cluster with less energy density towards the base station. This will ensure that energy is used more fairly among the sensor nodes, and the entire sensor network lives longer.

The rest of the paper is organized as follows. In Section 2, we detail the features of our grid network design and the cluster architecture, followed by the temperature conductivity and heat diffusion equation used in RETT in Section 3. Our experimental results are presented in Section 4. Finally, Section 5 presents our conclusions and future work.

2 Grid Network Design

We consider a wireless sensor grid network with N nodes placed in a two-dimensional area (x and y). This area is divided into small squares where each square covers an area of size A^2, where each node uses its location to associate itself with a point in the virtual grid. The number of squares will be related to the application conditions and dependent on its requirements. An example of how nodes determine their locations can be found in [11, 12, and 13]. The nodes will be self-organizing and not rely on deterministic deployment. When the nodes are scattered randomly, each group will form a square shaped cluster where every node in two adjacent clusters should be able to communicate directly. All nodes in the same square are grouped and each of them is distinguished using its own geographic address $(x_n, y_n, s, \tilde{n})$, where x_n, and y_n are its own coordinates, s is the cluster number and n is the sensor number within this particular cluster.

2.1 Cluster Architecture

Cluster heads are elected based on a temperature value corresponding to remaining energy that the sensor nodes have. The role of the cluster heads is to perform routing in the network. This will conserve energy since only one node at the time in each cluster is required to constantly operate with an active radio interface. The remaining sensors can resort to sensing the environment and only activate the radio when necessary. Heads of clusters will be changed once the current head of cluster temperature reaches a pre-determined threshold [15].

2.2 Clusters-Head Model

The main role for sensors is to sense the environment and to report the sensed data to a data sink. When a sensor has data to transmit, it will signal its cluster-head with the following information [15]:

- The sensor location.
- The sensed data and possibly additional information such as its type, value, position, and time.
- A timestamp indicating the urgency of the data, which is application dependent.

3 Equalization Based Routing

3.1 Heat Diffusion Equation

In RETT, we distribute our routing decisions based on the expected life span of nodes in the network. We transform this expected life span into a corresponding temperature and let each node make its routing decisions based on its surrounding temperatures. Thus, we wish to determine the temperature distribution in the network and from this information compute the conduction heat flux at any given point in the network using Fourier's Law [9]. The following equation is the general equation for heat diffusion [15].

$$T_{m,n+1} + T_{m,n-1} + T_{m+1,n} + T_{m-1,n} + \frac{q(\Delta x)^2}{\tau} - 4T_{m,n} = 0 \qquad (1)$$

3.2 Cluster-Head Temperature Calculation

Calculating the cluster-head temperature is divided into two phases; the first phase is to determine the direction of the routes between the clusters-heads and the base station (analogous to a constant gravity, pulling the data in the general direction of the sink), and the second phase is to select a suitable path or route that the cluster-head (the source) should use to send its data towards the base station on a per-packet and per-hop basis.

Once the sensor network has been established, each sink or base station should notify all the cluster-heads in the network about its location and the proposed values of each cluster-head temperature. The data will be saved within a table inside each cluster-head where each head saves its own temperature value in addition to temperature values of all neighbours found within the range of (x±1, y±1) from its location (x, y).

The mechanism for the first phase is as follows: Using the finite-difference equation, the base station will determine the proposed temperature for each cluster-head T_b by assuming the following values:

1- the power of each cluster-head will be set to zero $(q=0)$.

2- the base station will assume that its temperature will be the hottest point (i.e. $100C^o$) and the other boundary sides of the network area will be equal to $(0C^o)$. These two conditions in combination with the base station address will guarantee that the average direction of routing decisions in the network will be towards the base station and is done so that routing nodes can make fully distributed per-hop decisions rather than having to burden the network with defining pre-determined end to end routing paths.

Thus, if we consider a wireless sensor network of M finite-difference equations corresponding to M clusters whose temperatures are known. Identifying the clusters controllers by a single integer subscript (cluster number), rather than by the double subscript (m,n), we can determine the temperature distribution by solving these equations using an iterative numerical method such as *Gauss-Seidal method* [14].

The second phase follows the boot strapping of the network and is concerned with a per hop routing decision for each packet. The cluster-head (the source) will request the temperature and address of its neighbours stored in its data table. Upon receipt of this data it merges the temperature with the offset calculated in phase one and chooses the best neighbour based on the following three parameters:

1- The highest temperature value of a neighbour proposed by the base station T_b.

2- The highest real temperature value of a neighbour T_n.

3- Threshold value for the real temperature T_r.

The mechanism for selecting the next hop neighbour is achieved by:

   a) Selecting the neighbour cluster with the highest T_b.

   b) Comparing the real temperature value of the cluster with T_r.

   c) If the real temperature value is greater or equal to T_r, the cluster head will select this neighbour as a next hop.

   d) If the real temperature value is less than T_r, the cluster head will choose the neighbour with highest T_n as a next hop.

Each cluster will follow the same procedure to build up the entire path to the desired base station.

In the routing algorithm, we approximate the network with a square grid. The cluster head will firstly restrict the path for a packet by making sure that it will not move away from the base station, thus avoiding routing loops. After this boundary condition has been set, the node continues by selecting the hottest valid neighbouring node as the next hop node. This determined node will be selected for each packet until the routing path is re-calculated again according to the routing update period.

3.3 Complexity Analysis

The Gauss–Seidal Method introduced in the previous section, is suitable for solving M finite-difference equations corresponding to M clusters-heads with unknown temperature. In this section, we will show why this choice is acceptable by analysing the complexity of the method used in our model in terms of space complexity.

The Gauss-Seidal Method has been used in phase one for determining the direction of the routes between the cluster-heads and the base station and this calculation needs to be done only once, as long as the geometry does not change. Although this stage might require considerable time, the fact that it has to be done in the base station side and only once, makes this calculation mostly interesting in terms of memory consumption estimates. However, in practice, the number of iterations is proportional to the number of cluster-heads. Thus the memory space complexity in our model is O(n), where n is the number of clusters-heads.

4 Simulation Results

In our simulation studies, we used a 100-node network where nodes were randomly scattered between (x = 0, y = 0), and (x = 100, y =100) with one base station at location (x = 110, y =50). After generating events randomly, each data packet used to report these events was 500 bytes long, and for the purpose of this study, we used a similar energy model to the one proposed in [3].

We analyse the performance of RETT using a simulator written in MATLAB. In this paper, we consider a static and homogeneous sensor network, with a single data sink. In such a network, it is sufficient to determine geographical information at network formation only since nodes are static. We suppose that nodes are deployed randomly forming a high-density network in a flat topology. As a data delivery model, we simulate an event-driven network in which sensors report information only if an event of interest occurs. In this case, the base station is interested only in the occurrence of a specific set of events. Each cluster is set to 20m X 20m; hence there are a total of 25 clusters in the network.

In our simulations, we assume each node begins with 0.5J of energy and an unlimited amount of data generated in random fashion. For these simulations, energy is consumed whenever a node transmits or receives data or performs data aggregation. In order to evaluate the performance of RETT, we compare the protocol operating in Highest Energy (RETT-HE) and Equalizing Energy (RETT-EE) modes with the well-known and well-referenced LEACH [3] protocol. Moreover, we get each of the following results after running the simulator more than 10 times per experiment.

In RETT-HE mode; each cluster determines its next-hop neighbour solely depending on which of its neighbours has got the highest energy towards the base station. In RETT-EE mode, RETT tries to find the next-hop neighbour after comparing the neighbouring cluster-head's energy with a specified threshold determined by the user. If the energy is greater than the threshold it will select the shortest path mode and otherwise the RETT-HE mode will be chosen. In RETT-EE mode, we assume that cluster heads always transmit packets directly to the base station if the energy value of their neighbours are less than or equal to the threshold value. Hence, this constraint will allow the neighbours to only send their own packets towards the base station. Under the various used modes, the energy efficiency and the network lifetime have been evaluated. The same environment is employed in the simulations using different routing modes or the LEACH algorithm.

Fig. 1 shows the sum of the residual energies of all clusters in the network as a function of generated packets for the two modes using RETT and LEACH. The simulation results indicate that RETT-EE mode consumes less energy than both RETT-HE mode and the LEACH protocol while generating the same number of packets. In LEACH, nodes in the cluster only send data to their cluster head, which then sends its aggregated data packets directly to the base station over a long-range radio and hence might consume more energy for large-scale networks than using a multi-hop routing mechanism such as RETT. As a result, we can see that RETT-EE mode has increased the number of delivered packets when compared with RETT-HE mode and LEACH algorithm. In addition, the network shows an extended life span using RETT with limited coverage compared to LEACH.

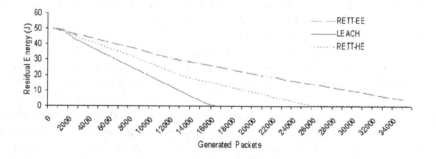

Fig. 1. Energy consumption with number of generated packets

The number of alive sensor nodes is shown in Fig. 2. Our simulation results show that the life time for sensor nodes using RETT with RETT-EE mode is significantly increased and hence yields longer stability for the sensor network. The reason is that the cluster heads always transmit packets directly to the base station if the energy value of their neighbors are less than or equal to the proposed threshold. Therefore, the constraint allows the neighbors to only send their own packets towards the base station. On the other hand, LEACH has got a high number of dead nodes within a small period of time and as a result, the cluster head election process becomes unstable and fewer nodes become cluster head. Even worse, during the last rounds, there are only few rounds when more than one cluster head is elected.

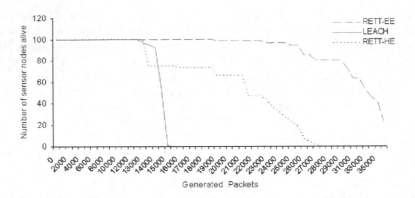

Fig. 2. Energy consumption with number of generated packets

5 Conclusions and Future Work

In order to maximize efficient usage of batteries used by randomly placed sensors, we proposed RETT; a scalable, energy-efficient clustering based routing protocol that aims at maximizing the life span of an entire sensor network by means of power equalization. The protocol is based on a heat conductivity analogy to provide fully distributed and autonomous routing decisions, thereby avoiding complex and costly management functions. RETT is completely distributed, requiring no control information from the base station, and the nodes do not require knowledge of the global network in order for RETT to operate.

To evaluate the performance of RETT, we have compared the protocol operating in two modes: Highest Energy and Equalizing Energy modes with LEACH algorithm. Based on experiments with our MATLAB simulator for RETT, Equalizing the energy among the nodes in the network is effective in maximizing the system lifetime from a global perspective.

Our future work will initially focus on extending the model to heterogeneous compositions of sensor nodes with less regularly-shaped clusters and to investigate how RETT will perform under such conditions. We also intend to conduct an analysis to determine how the predictability of node failure is affected by the heterogeneity. Furthermore, we will endeavour to study RETT's performance under node mobility since mobility brings new challenges in equalization, as it is counter active to power equalization.

References

1. Akyildiz I., Su, W., Sankarasubramaniam, Y., Cayirei, E.: Wireless sensor networks: a survey. Computer Networks, Vol. 38 (2002) 393-422
2. Zhao, F., Guibas, L.: Wireless Sensor Networks: An Information Processing Approach. Morgan Kaufmann Publishers, California (2004)

3. Heinzelman, W., Chandrakasan, A., Balakrishnan, H.: Energy-Efficient Communication Protocol for Wireless Microsensor Networks. In HICSS'33 (2000) 3005-3014
4. Singh, S., Woo, M., Raghavendra, C.: Power-aware routing in mobile ad hoc networks, In: ACM/IEEE MOBICOM (1998) 181-190
5. Chang, J.-H., Tassiulas, L.: Maximum Lifetime Routing in Wireless Sensor Networks. IEEE/ACM Transaction on Networking, Vol. 12 (2004) 609-619
6. Xu, Y., Heidemann, J., Estrin, D.: Geography-informed Energy Conservation for Ad Hoc Routing. In ACM MOBICOM'01 (2001)
7. Schurgers, C., Tsiatsis, V., Ganeriwal, S. Srivastava, M.: Topology Management for Sensor Networks: Exploiting Latency and Density. In ACM MOBIHOC'02 (2002)
8. Misra, A., Banerjee, S.: MRPC: maximizing network lifetime for reliable routing in wireless Environments. In IEEE Wireless Communications and Networking Conference (WCNC), Vol. 2 (2002) 800-806
9. Chandrupatla, T., Belegundu, A.: Introduction to Finite Elements in Engineering. 3rd edn. Prentice-Hall Inc., New Jersey (2002)
10. Middleman, S.: An Introduction to Mass and Heat Transfer: Principles of Analysis and Design. John Wiley & Sons Inc., New York (1998)
11. Savvides, A., Han, C., Strivastava, M.: Dynamic fine-grained localization in ad-hoc networks of sensors. In ACM MOBICOM'01 (2001) 166-179
12. He, T., Huang, C., Blum, B., Stankovic, J., Abdelzaher, T.: Range-free localization schemes for large scale sensor networks. In MOBICOM'03 (2003) 81-95
13. Nasipuri, A., Li, K.: A directionality based location discovery scheme for wireless sensor networks. In the first ACM international workshop on Wireless Sensor Networks and Applications (2002) 105-111
14. Jordan, D., Smith, P.: Mathematical Techniques. Oxford University Press Inc. New York (1994)
15. Matrouk, K., Landfeldt, B.: Energy-Conservation Clustering Protocol based on Heat Conductivity for Wireless Sensor Networks. In IEEE ISSNIP (2004) 19-24

On the Information Flow Required for Tracking Control in Networks of Mobile Sensing Agents*

Sandip Roy, Liang Chen, and Ali Saberi

Washington State University

Abstract. We design controllers that permit a network of mobile agents with distributed sensing capabilities to track (follow) desired trajectories, and identify what trajectory information must be distributed to each agent for tracking.

1 Introduction

In several modern applications, teams of autonomous agents with distributed sensing and/or communication capabilities are required to cooperatively complete a complex task (see, e.g., [1,2,3]). In many of these applications, the teams must complete *tracking tasks*—ones in which each agent's state (e.g., position) must follow a specified command signal (e.g. [3,4,5]). For instance, teams of autonomous vehicles which sense relative positions may need to follow a "lawn-mower" pattern, to search a minefield. Similarly, a bank of antennas may need to follow a path in a coordinated fashion. Very broadly, these tracking problems require two sorts of *information flow*: 1) local sensing for *control*, i.e. using which each agent can correctly actuate its dynamics (change its state) so as to follow a specified path; and 2) higher-level communication for *trajectory-distribution*, i.e. for disseminating each agent's desired path to it. Often, in tracking applications for modern communicating-agent teams, the information flow must be highly limited, in that it must be both local in space and sparse in time (see, e.g., [5] for motivation regarding AUVs). In this article, we marry well-known techniques for servo control with techniques for decentralized control and distributed-algorithm-development, to develop low-information-flow tracking algorithms for networks of sensing/communicating agents. In doing so, we delineate the role played by sensing topology of the network in our ability to achieve tracking. We also study what trajectory information must be distributed to the agents, and explore means for distributing this information with sparse communication.

Recently, several articles have sought to expose the role played by the sensing topology for *formation* (convergence to a fixed or constant-velocity pattern) in communicating-agent networks (e.g., [1,2]). Our recent work on *formation* has exposed that linear decentralized static and dynamic controllers can be

---

* Correspondence should be sent to the first author, at sroy@eecs.wsu.edu. This work was partially supported by the National Science Foundation under Grant ECS 0528882 (Sensors), and by the Office of Naval Research under Grant N000140310848.

S. Nikoletseas and J.D.P. Rolim (Eds.): ALGOSENSORS 2006, LNCS 4240, pp. 208–216, 2006.

used to stabilize a network of double-integrator agents with sensing capabilities and actuator saturation, under broad connectivity requirements on the sensing topology [2]. Our aim here is to extend the controllers/algorithms for formation in double-integrator networks to permit tracking of temporal signals rather than simple approach to a fixed value, for general communication/sensing topologies.

Our work connects with the classical literature on decentralized tracking (e.g [6]), as well as more recent studies on tracking control for (specifically) teams of autonomous vehicles (e.g. [4,5]). With respect to the classical literature, our study differs significantly in the following sense: our class of decentralized control problems—which are appropriate for sensing-network applications—requires that trajectories from a set of neighbors (or, more specifically, a function of these trajectories) be distributed to each agent; sending only the desired trajectory for a particular agent to that agent is insufficient. With regard to the recent literature on autonomous-vehicle control, our approach differs in that it permits us to decouple the stabilization, path-following, and path-distribution aspects of the control task. Hence, we are able to achieve tracking for a very general class of sensing topologies, and to track essentially arbitrary signals when sufficient communication for path-distribution is permitted.

2   Model and Problem Formulation

In this section, we motivate and describe a model comprising a network of communicating/sensing agents with double-integrator internal dynamics, and then introduce the *tracking problem* in the context of this *double-integrator network*.

2.1   The Double Integrator Network

We consider a network of n mobile sensors or vehicles or **agents**, labeled $1, \ldots, n$. These agents aim to cooperatively complete a dynamic task, by actuating their internal dynamics using sensed observations. Let us motivate and introduce our model for the internal dynamics for the agents, and then consider the sensing architecture.

Increasingly, modern *communicating-agent networks* (networks of mobile sensors, autonomous vehicle teams) are made up of agents with simple but highly-constrained dynamics. For many such networks, a plausible canonical model for an agent's dynamics is a saturating double integrator. We adopt this saturating double-integrator model for our agents' internal dynamics, i.e., we assume agent i is governed by $\ddot{r}_i = \sigma(u_i)$, where r_i is the **position**[1] of agent i, u_i is agent i's input or actuation, and $\sigma()$ is the standard saturation function. We also find it convenient to refer to the derivative of agent i's position with respect to time,

---

[1] We use the term **position** for an agent's state with the vehicle-control application in mind (in which Newton's law yields the double-integrator model), but in fact double-integrator models can capture various dynamics of interest in communicating-agent networks (see [2] for some examples).

i.e. $v_i = \dot{r}$) as agent i's velocity. We note that for the above dynamics, each agent is assumed to have a scalar position. However, there is no loss of generality since it turns out that the internal dynamics, observations and controller of a multiple-dimensional agent can be equivalently represented using multiple scalar agents (see [2]).

For some applications (e.g., computational ones, where agents' states represent opinions/estimates rather than actual positions), actuator saturation may not be a significant constraint. Motivated by such applications, and so as to make the presentation of results clearer, we often first consider the case where agents are governed by pure double integrators $\ddot{r}_i = u_i$ rather than saturating double integrators.

Each agent in the network has certain sensing capabilities. Our model for sensing is quite general: each agent is assumed to make multiple observations, each of which is a linear combination of (in general multiple) agents' current positions or velocities. We note that such a model permits representation of absolute and relative position observations, among others. Specifically, we suppose agent i has $2m_i$ observations on the network (m_i observations on positions, and m_i observations on velocities) that can be written in the form $\mathbf{a}_{p_i} = G_i \mathbf{r}$ and $\mathbf{a}_{v_i} = G_i \mathbf{v}$ respectively, where $\mathbf{r}^T \triangleq [r_1 \dots r_n]$, $\mathbf{v}^T \triangleq [v_1 \dots v_n]$, and the **graph matrix** G_i has dimension $m_i \times n$. We note that each agent is assumed to have identically-structured observations on positions and velocities. This restriction is plausible in many applications (see [2] for motivation) and also can be eliminated by considering more general controllers than pursued here; the methods developed here can be generalized to this case. For convenience, we also define the **full graph matrix** as $G = \begin{bmatrix} G_1^T \dots G_n^T \end{bmatrix}^T$.

We assume that agent i has available its observations for computing its actuation (input). That is, the observations $\mathbf{a}_{p_i}$ and $\mathbf{a}_{v_i}$ of agent i are considered as information that is available to the local controller. Our goal is to design a *static linear controller* (a controller without memory, specifically one that sets the input u_i to a linear combination of the current observations) for each agent i, so as to globally achieve a tracking task[2].

We refer to the internal dynamics, communication topology, and decentralized control paradigm described above together as a **saturating double-integrator network** (respectively, **double integrator network** in the case where actuator saturation is ignored).

2.2   Tracking Problem Formulation

Motivated by a range of applications, we aim to design controllers for the double-integrator network for **tracking**, i.e. controllers using which each agent i's position can follow a **desired trajectory** $\bar{r}_i(t)$, $t \geq 0$. We find it convenient to refer

---

[2] Our motivation for consider static controllers is that they are easily implementable even in devices with limited complexity. Consideration of static control also clarifies the exposition of information flow for tracking; many results readily generalize to settings where more complicated controllers are used.

to a set of desired trajectories $\bar{r}_1(t), \ldots, \bar{r}_n(t)$ together as the **tracking task** $(\bar{r}_1, \ldots, \bar{r}_n)$.

Typically, in the controls literature, desired trajectories are assumed to be ones that can be generated by an autonomous linear system (which is termed an *exosystem*): almost all trajectories of interest (for instance, ramp, step, or sinusoidal signals) can be represented in this way (see e.g. [7]). For us, signals of this form are compelling with respect to their distribution to agents, since they can be specified by the parameters (more specifically, *eigenvalues* or *modes*) of the autonomous systems and the initial condition of the autonomous system's state. Thus, whenever convenient, we assume (without meaningful loss of generality) that the each desired trajectory is generated by the exosystem $\dot{\mathbf{w}}_i = S_i \mathbf{w}_i$, $\bar{r}_i = \mathbf{d}_i^T \mathbf{w}_i$, where $\mathbf{w}_i \in R^{q_i}$, the system matrix S_i is assumed without loss of generality to have eigenvalues in the CRHP (since tracking is an asymptotic task, see [7]), and the internal conditions $\mathbf{w}_i(0)$ are set so that the desired trajectory is generated. We kindly refer the reader to the extended version of our work in [9] for an illustration of trajectory-generation from an exosystem.

Our aim is to design controllers for each agent, so that their positions follow the desired trajectories in an asymptotic sense. Let us define the achievement of the tracking task formally:

Definition 1. *A double-integrator network is said to achieve or complete the tracking task* $(\bar{r}_1, \ldots, \bar{r}_n)$, *if the error signals* $e_i(t) \triangleq r_i(t) - \bar{r}_i(t)$, $i = 1, \ldots, n$, *approach 0 as* $t \to \infty$.

We note that tracking is an *asymptotic* task; a settling time describing how quickly tracking is achieved can be obtained, see [7] for details.

3 Tracking and the Required Information Flow

In this section, we first give conditions under which a double-integrator network (with and without actuation saturation) can achieve tracking and show how a controller can be designed to do so. Throughout this section, we assume full information flow regarding the tracking task, i.e., we assume that the agents are given all required information about the signals in the tracking task $(\bar{r}_1, \ldots, \bar{r}_n)$. We also delineate carefully what information about the tracking task must be given to each agent.

First, we consider a double-integrator network that is not subject to actuation saturation, and show how to develop a controller for the tracking task, in the process giving broad conditions under which tracking is possible.

Theorem 1. *A double-integrator network can complete any tracking task* $(\bar{r}_1, \ldots, \bar{r}_n)$ *if there exists a block diagonal matrix* $K = diag[\mathbf{k}_i^T]$ *(where each* $\mathbf{k}_i$, $i = 1, \ldots, n$ *is an* m_i-*component vector) such that all eigenvalues of* KG *are in the OLHP.*

In this case, the tracking task can be achieved by driving each agent i *with the input*

$$u_i = K \left(\begin{bmatrix} \mathbf{a}_{p_i} \\ \mathbf{a}_{v_i} \end{bmatrix} - \begin{bmatrix} G_i D \mathbf{w} \\ G_i D S \mathbf{w} \end{bmatrix} \right) + \mathbf{d}_i^T S_i^2 \mathbf{w_i},$$ *where α is a sufficiently large positive number,* $D = diag[\mathbf{d}_i^T]$, $S = diag[S_i]$, *and* $\mathbf{w}^T = \begin{bmatrix} \mathbf{w}_1^T \dots \mathbf{w}_n^T \end{bmatrix}$.

We refer the reader to the extended version of our work in [9] for a proof of this key theorem. A couple notes about this theorem are worthwhile:

1) We stress that the controllers being used to achieve the tracking task are *static* (memoryless) linear ones: the input $u_i(t)$ at each time t is a linear function of the current observations.

2) We have chosen to present the control inputs in terms of the exosystem parameters, with the motivation that this form is often the most easily implemented one. We can easily phrase the input in terms of the desired trajectories. In this notation, we find that $u_i = K \left(\begin{bmatrix} \mathbf{a}_{p_i} \\ \mathbf{a}_{v_i} \end{bmatrix} - \begin{bmatrix} G_i \bar{\mathbf{r}} \\ G_i \dot{\bar{\mathbf{r}}} \end{bmatrix} \right) + \ddot{\bar{r}}_i$, where $\bar{\mathbf{r}}^T = \begin{bmatrix} \bar{r}_1 \dots \bar{r}_n \end{bmatrix}$.

According to Theorem 1, tracking is possible whenever there exists an appropriately-structured (block-diagonal) K such that KG has all eigenvalues in the OLHP. Essentially, whenever this condition holds, the controller can be chosen to make the stable[3]. Thus, we refer a double-integrator network for which there is K such that KG has all eigenvalues in the OLHP as a **stabilizable double-integrator network**. We note that the linear-algebra problem of whether there exists block-diagonal K such that KG has eigenvalues in the OLHP is well-studied (e.g., [8]), and there are several broad classes of full graph matrices for which such stabilizing K exist. We refer the reader to [2] for details.

So far, we have developed conditions under which an unsaturating double-integrator network can achieve tracking; we note that the amplitude of the input signal u_i may be arbitrarily large when the controller in Theorem 1 is used, and hence the result does not necessarily carry through to the case where actuators may saturate. We next develop a condition for tracking in a saturating double-integrator network. Conceptually, tracking under saturation requires the further condition that the actuator can provide enough acceleration at all times to move each agent along its desired trajectory, plus an arbitrarily small amount of further acceleration for convergence to the trajectory (stabilization). Under these conditions, by making the convergence to the trajectory sufficiently slow, we can achieve tracking for an arbitrarily large set of initial conditions. That is, tracking is achieved in a *semi-global* sense, i.e. given any closed and bounded ball of initial conditions for the saturating double-integrator network, there exists a controller that achieves tracking for any initial condition in this ball. This notion is formalized in the following theorem:

Theorem 2. *Consider the tracking problem of a double-integrator network with n agents that are subject to actuation saturation. A tracking task $(\bar{r}_1, \dots, \bar{r}_n)$ can be achieved for any given closed and bounded set of initial conditions, if the two following conditions hold:*

---

[3] That is, the desired trajectories are attractive and also stable in the sense of Lyapunov.

(I) There exists a block diagonal matrix $K = diag([\mathbf{k}_i^T])$ (where $\mathbf{k}_i$ is an m_i-component vector) such that all eigenvalues of KG are in the OLHP.
(II) There exists a $\delta > 0$ and a $T \geq 0$ such that $\|\mathbf{d}_i^T S_i^2 \mathbf{w}_i\|_{\infty,T} \leq 1 - \delta$, (or equivalently, $\|\frac{d^2\bar{r}_i}{dt^2}\|_{\infty,T} \leq 1 - \delta$) for all $i = 1, \ldots, n$.

We refer the reader to [9] for the proof of this theorem. Illustrations of the tracking algorithms can also be found in [9].

From the form of the inputs in Theorems 1 and 2, we notice that each agent i's input depends only on its own observations, as required; however, computation of the input in general requires knowledge of the desired trajectories of multiple agents. In the subsequent sections, we will develop methods for providing the agents with this information using minimal communication when the agents are known to move in formation, and will even explore whether agents can deduce the required trajectories when only a leader has been told the tracking task. Before considering these special cases, we first identify the trajectory information that in general must be provided to agent i to achieve the tracking task. We find that the required information is deeply connected to the structure of the communication/sensing topology, i.e., of the graph matrices $G_1, \ldots, G_n$. To present this result, let us formulate the notion of a **graph** for the network communication/sensing:

Definition 2. *The **network graph** is a directed graph with n nodes labeled $1, \ldots, n$, which corresponds to the n agents. The network graph has an edge from vertex j to vertex i, if and only if $G_{ij} \neq \mathbf{0}$, where G_{ij} is the jth column of the graph matrix G_i.*

We note that a directed edge from vertex j to vertex i indicates that the observations made by agent i depend on the position/velocity of agent j[4].

We also find it convenient to define the notion of neighbors from the network graph. In particular, if the network graph has an edge from vertex j to vertex i, we refer to vertex (equivalently, agent) j as an **upstream neighbor** of vertex (equivalently, agent) i. We use the notation U(i) for the set of upstream neighbors of agent i, and use the term **upstream neighbor set** for this set.

Let us now give the general result on the trajectory information required by each agent:

Theorem 3. *Each agent i requires a signal $\bar{z}_i$, which is a function of the trajectories $\bar{r}_j$, $j \in \mathcal{U}(i)$, as well as the trajectory $\bar{r}_i$, in order that the tracking task can be achieved. Specifically, the agent i requires the signal $\bar{z}_i(\bar{r}_{\mathcal{U}(i)}, \bar{r}_i) = -K_i^T \Sigma_{j \in \mathcal{U}(i)} G_{ij}(\bar{r}_j + \alpha\frac{d\bar{r}_j}{dt}) + \frac{d^2\bar{r}_i}{dt^2}$, where we have used the notation $\bar{r}_{\mathcal{U}(i)}$ for the set of trajectories $\bar{r}_j$, $j \in \mathcal{U}(i)$.*

Theorem 3 makes clear that each agent i requires a signal $\bar{z}_i$, which is a function of the trajectories that the agent and its upstream neighbors must follow. Since

---

[4] We permit *self-loops*, i.e. edges from a node back to itself. A self-loop at vertex i indicates that the observations made by agent i depend on its own position and velocity.

the statistic $\bar{z}_i(\bar{r}_{\mathcal{U}(i)}, \bar{r}_i)$ specifies the trajectory information that must be distributed to or computed by agent i, we refer to the function $\bar{z}_i()$ as the **trajectory information distribution function (TIDF)** for agent i, and refer to a signal $\bar{z}_i$ generated by this function for a particular set of trajectories as a **trajectory information signal (TIS)** for agent i.

We stress here that the tracking task CANNOT be achieved by simply sending each agent its own desired trajectory: the TIS for agent i depends on the desired trajectories not only of agent i but of its upstream neighbors. More specifically, agent i's input (actuation) can be viewed as comprising two components, one $(\dfrac{d^2\bar{r}_i}{dt^2})$ that provides the agent with the power needed to follow its desired trajectory and only depends on its own trajectory, and one $(-K_i^T \Sigma_{j\in\mathcal{U}(i)} G_{ij}(\bar{r}_j + \alpha\dfrac{d\bar{r}_j}{dt}))$ that is needed for agents to stabilize about the trajectory and depends on the desired trajectories of upstream neighbors. We refer the readers to an extended version of this work [9] for a discussion on the distribution of trajectory information.

We have thus given conditions on the communication/sensing topology for tracking in double integrator networks and saturating double integrator networks. In the process, we have exposed the need for trajectory-distribution, identified the trajectory information signals needed by each agent, and indicated the dependence of these signals on the sensing/communication topology.

4   Information Distribution in Formation

For many autonomous vehicle control and mobile sensing applications, the agents in a network are required to move in *formation*, i.e. they maintain a geometric pattern as they move through space. To consider tracking in formation, we must impose a geometric (spatial) interpretation for the agents' states. Specifically, motivated by typical autonomous-vehicle-control applications, we consider a network of n agents moving in the plane. We refer to such a model as a **planar double integrator network (PDIN)**, see [9] for details.

Our goal is to have each agent in a PDIN follow a **desired x-trajectory** (desired trajectory in the x-direction) $\bar{r}_{ix}(t)$ and **desired y-trajectory** $\bar{r}_{iy}(t)$. This **tracking task** is a formation-tracking task if the agents maintain a rigid pattern as they move, as formalized below.

Definition 3. *A nominal formation F_0 is an ordered set of n pairs $(\widehat{r}_{1x}, \widehat{r}_{1y})$, $\ldots, (\widehat{r}_{nx}, \widehat{r}_{ny})$, along with a reference $\widehat{r}_{0x}$ and $\widehat{r}_{0y}$. The nominal formation describes a pattern of points in the plane, together with a reference point for this pattern.*

Definition 4. *An ordered set of n points in the plane (i.e., n pairs) (r_{1x}, r_{1y}), $\ldots, (r_{nx}, r_{ny})$ is said to be in the formation F_0, if all points $(\widehat{r}_{ix}, \widehat{r}_{iy})$ in the nominal formation can be placed on the corresponding points (r_{ix}, r_{iy}) through expansion around the reference point, rotation around the reference point, and* *translation in the x- and y- directions. That is,* $(r_{1x}, r_{1y}), \ldots, (r_{nx}, r_{ny})$ *is in formation* F_0 *if there are parameters* a, θ, p_x, *and* p_y *such that* $r_{ix} = a[(\hat{r}_{ix} - \hat{r}_{0x})cos(\theta) + (\hat{r}_{iy} - \hat{r}_{0y})sin(\theta)] + p_x$ *and* $r_{iy} = a[(\hat{r}_{ix} - \hat{r}_{0x})sin(\theta) + (\hat{r}_{iy} - \hat{r}_{0y})cos(\theta)] + p_y$, *for all* i.

Definition 5. *A formation-F_0 tracking task is one in which the set of desired trajectories* $(\overline{r}_{1x}(t), \overline{r}_{1y}(t)), \ldots, (\overline{r}_{nx}(t), \overline{r}_{ny}(t))$ *is in the formation* F_0, *at each time t.*

We have been able to show that formation-tracking can in many cases be achieved with less information flow than required for arbitrary tracking. We omit these results in the interest of space (see [9] for details), and simply present a couple simulations of formation-tracking tasks (Figure 1).

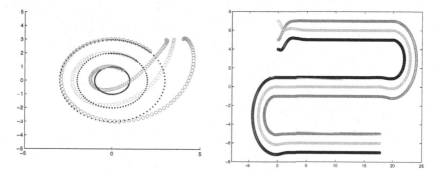

Fig. 1. The left figure shows a circular in-formation tracking, and the right figure shows an in-formation tracking of a lawn-mower pattern (there are two formations, since two different reference points are used when making the two turns)

References

1. J. A. Fax and R. M. Murray, "Information flow and cooperative control of vehicle formations", submitted to *IEEE Transactions on Automatic Control*, April 2003.
2. S. Roy, A. Saberi, and K. Herlugson, 'Formation and Alignment of Distributed Sensing Agents with Double-Integrator Dynamics', IEEE Press Monograph on *Sensor Network Operations*, 2004 (in press).
3. A. Williams, G. Lafferriere, and J. J. P. Veerman, "Stable motions of vehicle formations," in *Proceedings of the 2005 Conference on Decision and Control.*
4. P. Ogren, E. Fiorelli, and N. Leonard, "Cooperative control of mobile sensor networks," *IEEE Transactions on Automatic Control*, vol. 49, no. 8, 2004.
5. D. J. Stilwell, B. E. Bishop, and C. A. Sylvester, "Redundant manipulator techniques for partially decentralized path planning and control of a platoon of autonomous vehicles," *IEEE Transactions on Systems, Man, and Cybernetics*, vol. 35, no. 4, Aug. 2005.
6. E. Davison, "The robust decentralized control of a general servomechanism problem," *IEEE Transactions on Automatic Control*, vol. 21, no. 1, pp. 14-24, Feb. 1976.

7. A. Saberi, A. Stoorvogel, and P. Sannuti, *Output Regulation and Control Problems with Regulation Constraints*, Springer-Verlag, 1999.
8. M. E. Fisher and A. T. Fuller, "On the stabilization of matrices and the convergence of linear iterative processes," *Proceedings of the Cambridge Philosophical Society*, pp. 417–425, 1956.
9. L. Chen, S. Roy, and A. Saberi, "On the Information Flow Required for Tracking Control in Networks of Mobile Sensing Agents," submitted to the *IEEE Transactions on Mobile Computing*.

Author Index

Lecture Notes in Computer Science

For information about Vols. 1–4263

please contact your bookseller or Springer